T0297596

The Designer's Guide to the Cortex-M Processor Family

To my wife Sarah and my parents Ann and Maurice

The Designer's Guide to the Cortex-M Processor Family

A Tutorial Approach

Trevor Martin

AMSTERDAM • BOSTON • HEIDELBERG • LONDON
NEW YORK • OXFORD • PARIS • SAN DIEGO
SAN FRANCISCO • SINGAPORE • SYDNEY • TOKYO
Newnes is an imprint of Elsevier

Newnes is an imprint of Elsevier

The Boulevard, Langford Lane, Kidlington, Oxford, OX5 1GB

225 Wyman Street, Waltham, MA 02451, USA

Notice

Knowledge and best practice in this field are constantly changing. As new research and experience
broaden our understanding, changes in research methods, professional practices, or medical treatment may
become necessary.

Practitioners and researchers must always rely on their own experience and knowledge in evaluating
and using any information, methods, compounds, or experiments described herein. In using such information
or methods they should be mindful of their own safety and the safety of others, including parties for whom
they have a professional responsibility.

To the fullest extent of the law, neither the Publisher nor the authors, contributors, or editors, assume any
liability for any injury and/or damage to persons or property as a matter of products liability, negligence
or otherwise, or from any use or operation of any methods, products, instructions, or ideas contained in
the material herein.

British Library Cataloguing-in-Publication Data

A catalogue record for this book is available from the British Library

Library of Congress Cataloging-in-Publication Data

A catalog record for this book is availabe from the Library of Congress

ISBN: 978-0-08-098296-0

For information on all Newnes publications
visit our website at store.elsevier.com

Typeset by MPS Limited, Chennai, India
www.adi-mps.com

Printed and bound in the United States

13 14 15 16   10 9 8 7 6 5 4 3 2 1

Contents

Foreword

Today, there is a silent revolution in the embedded market. Most new microcontrollers and application processors that are introduced are based on the ARM® architecture. Recently, we also saw the launch of several new ARM processors. At the low end of the spectrum, the Cortex™-M0+ processor has been introduced for applications which were previously dominated by 8-bit microcontrollers. The new 64-bit Cortex-A50 series processors address the high-end market such as servers. Gartner is forecasting 50 billion devices that are connected to the Internet-of-Things (IoT) in the year 2020 and ARM processors span already the whole application range from sensors to servers. Many of the sensor devices will be based on Cortex-M microcontrollers and use just a small battery or even energy harvesting as power source.

With the availability of even more capable microcontrollers, software development for these devices has become more complex over the years. Use of real-time operating systems is rapidly becoming an industry best practice, and the use of commercial middleware as well as reuse of custom libraries is gaining importance for cost-efficient software engineering. Successfully combining these building blocks of a modern embedded application often poses a problem for developers. Industry standards are a great way to reduce system development costs and speed up time to market. And the Cortex-M processor architecture along with the CMSIS software programming standard is the basis for further hardware and software standardization.

Reinhard Keil

Preface

ARM first introduced the Cortex-M processor family in 2004. Since then the Cortex-M processor has gained wide acceptance as a general purpose processor for small microcontrollers. At the time of writing there are over 1000 standard devices that feature the Cortex-M processor available from many leading semiconductor vendors and the pace of development shows no sign of slowing. While predicting the future is always a dangerous hobby, the Cortex-M processor is well on its way to becoming an industry standard architecture for embedded systems. As such the knowledge of how to use it is becoming a requisite skill for professional developers. This book is intended as both an introduction to the Cortex-M processor and a guide to the techniques used to develop application software to run on it. The book is written as a tutorial and the chapters are intended to be worked through in order. Each chapter contains a number of examples that present the key principles outlined in this book using a minimal amount of code. Each example is designed to be built with the evaluation version of the MDK-ARM. These examples are designed to run in a simulator so you can use this book without any additional hardware. That said examples can also be run on a number of low-cost hardware modules that are widely available through the Internet.

Chapter 1 provides an introduction and feature overview of each processor in the Cortex-M family.

Chapter 2 introduces you to the basics of building a C project for a Cortex-M processor.

Chapter 3 provides an architectural description of the Cortex-M3 and its differences to the other Cortex-M processors.

Chapter 4 introduces the CMSIS programming standard for Cortex-M processors.

Chapter 5 extends Chapter 3 by introducing the more advanced features of the Cortex-M architecture.

Chapter 6 introduces the use of an RTOS on a Cortex-M processor.

Chapter 7 looks at the math and DSP support available on the Cortex-M4 and how to design real-time DSP applications.

Chapter 8 provides a description of the CoreSight debug system and its real-time features.

This book is useful for students, beginners, and advanced and experienced developers alike. However, it is assumed that you have a basic knowledge of how to use microcontrollers and that you are familiar with the instruction set of your preferred microcontroller. In addition, it is helpful to have basic knowledge on how to use the μVision debugger and IDE.

Acknowledgments

I would like to thank Charlotte Kent and Tim Pitts of Elsevier and Joseph Yui and Richard York of ARM.

About the Author

Trevor Martin is Senior Technical Specialist within Hitex UK. Over the 20 years he has worked at Hitex UK, Trevor has worked with a wide range of microcontrollers and associated development tools. Since the launch of the Cortex-M3 processor in 2004, Trevor has contributed numerous articles and application notes for many of the leading Cortex-M-based microcontrollers. Having an extensive knowledge of the Cortex-M processor family, Trevor is also familiar with many of the development techniques, application software, and communication protocols required for today's embedded applications.

CHAPTER 1

Introduction to the Cortex-M Processor Family

Cortex Profiles

In 2004, ARM introduced its new Cortex family of processors. The Cortex processor family is subdivided into three different profiles. Each profile is optimized for different segments of embedded systems applications.

Figure 1.1

The Cortex processor family has three profiles—application, real time, and microcontroller.

The Cortex-A profile has been designed as a high-end application processor. Cortex-A processors are capable of running feature-rich operating systems such as WinRT and Linux. The key applications for Cortex-A are consumer electronics such as smart phones, tablet computers, and set-top boxes. The second Cortex profile is Cortex-R. This is the real-time profile that delivers a high-performance processor which is the heart of an application-specific device. Very often a Cortex-R processor forms part of a "system-on-chip" design that is focused on a specific task such as hard disk drive (HDD) control, automotive engine management, and medical devices. The final profile is Cortex-M or the microcontroller profile. Unlike earlier ARM CPUs, the Cortex-M processor family has been designed specifically for use within a small microcontroller. The Cortex-M processor currently comes in five variants: Cortex-M0, Cortex-M0+, Cortex-M1, Cortex-M3, and Cortex-M4. The Cortex-M0 and Cortex-M0+ are the smallest processors in the family. They allow silicon manufacturers to design low-cost, low-power devices that can replace existing 8-bit microcontrollers while still offering 32-bit performance. The Cortex-M1 has much of the same features as the Cortex-M0 but has been designed as a "soft core" to run inside an Field Programmable Gate Array (FPGA) device. The Cortex-M3 is the mainstay of the Cortex-M family and was the first Cortex-M variant to be launched. It has enabled a new generation of high-performance 32-bit microcontrollers which can be manufactured at a very low cost. Today, there are many Cortex-M3-based microcontrollers available from a wide variety of silicon manufacturers. This represents a seismic shift where Cortex-M-based microcontrollers are starting to replace the traditional 8\16-bit microcontrollers and even other 32-bit microcontrollers. The highest performing member of the Cortex-M family is the Cortex-M4. This has all the features of the Cortex-M3 and adds support for digital signal processing (DSP) and also includes hardware floating point support for single precision calculations.

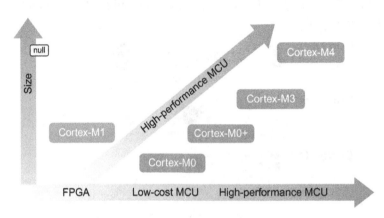

Figure 1.2
The Cortex-M profile has five different variants with a common programmers model.

In the late 1990s, various manufacturers produced microcontrollers based on the ARM7 and ARM9 CPUs. While these microcontrollers were a huge leap in performance and competed in price with existing 8/16-bit architectures, they were not always easy to use. A developer would first have to learn how to use the ARM CPU and then have to understand how a specific manufacturer had integrated the ARM CPU into their microcontroller system. If you have moved to another ARM-based microcontroller you might have gone through another learning curve of the microcontroller system before you could confidently start development. Cortex-M changes all that; it is a complete Microcontroller Unit (MCU) architecture, not just a CPU core. It provides a standardized bus interface, debug architecture, CPU core, interrupt structure, power control, and memory protection. More importantly, each Cortex-M processor is the same across all manufacturers, so once you have learned to use one

Cortex-M-based processor you can reuse this knowledge with any other manufacturers of Cortex-M microcontrollers. Also within the Cortex-M family, once you have learned the basics of how to use a Cortex-M3, then you can use this experience to develop using a Cortex-M0, Cortex-M0+, or a Cortex-M4 device. Through this book, we will use the Cortex-M3 as a reference device and then look at the differences between Cortex-M3 and Cortex-M0, Cortex-M0+, and Cortex-M4, so that you will have a practical knowledge of all the Cortex-M processors.

Cortex-M3

Today, the Cortex-M3 is the most widely used of all the Cortex-M processors. This is partly because it has been available not only for the longest period of time but also it meets the requirements for a general-purpose microcontroller. This typically means it has a good balance between high performance, low power consumption, and low cost.

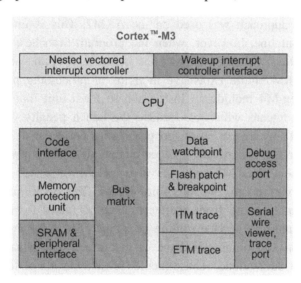

Figure 1.3
The Cortex-M3 was the first Cortex-M device available. It is a complete processor for a general-purpose microcontroller.

The heart of the Cortex-M3 is a high-performance 32-bit CPU. Like the ARM7, this is a reduced instruction set computer (RISC) processor where most instructions will execute in a single cycle.

Figure 1.4
The Cortex-M3 CPU has a three-stage pipeline with branch prediction.

This is partly made possible by a three-stage pipeline with separate fetch, decode, and execute units.

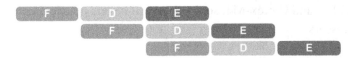

Figure 1.5
The Cortex-M3 CPU can execute most instructions in a single cycle. This is achieved by the pipeline executing one instruction, decoding the next, and fetching a third.

So while one instruction is being executed, a second is being decoded, and a third is being fetched. The same approach was used on the ARM7. This is great when the code is going in a straight line, however, when the program branches, the pipeline must be flushed and refilled with new instructions before execution can continue. This made branches on the ARM7 quite expensive in terms of processing power. However, the Cortex-M3 and Cortex-M4 include an instruction to fetch unit that can handle speculative branch target fetches which can reduce the bench penalty. This helps the Cortex-M3 and Cortex-M4 to have a sustained processing power of 1.25 DMIPS/MHz. In addition, the processor has a hardware integer math unit with hardware divide and single cycle multiply. The Cortex-M3 processor also includes a nested vector interrupt unit (NVIC) that can service up to 240 interrupt sources. The NVIC provides fast deterministic interrupt handling and from an interrupt being raised to reaching the first line of "C" in the interrupt service routine takes just 12 cycles every time. The NVIC also contains a standard timer called the systick timer. This is a 24-bit countdown timer with an auto reload. This timer is present on all of the different Cortex-M processors. The systick timer is used to provide regular periodic interrupts. A typical use of this timer is to provide a timer tick for small footprint real-time operating systems (RTOS). We will have a look at such an RTOS in Chapter 6. Also next to the NVIC is the wakeup interrupt controller (WIC); this is a small area of the Cortex-M processor that is kept alive when the processor is in low-power mode. The WIC can use the interrupt signals from the microcontroller peripherals to wake up the Cortex-M processor from a low-power mode. The WIC can be implemented in various ways and in some cases does not require a clock to function; also, it can be in a separate power region from the main Cortex-M processor. This allows 99% of the Cortex-M processor to be placed in a low-power mode with just minimal current being used by the WIC.

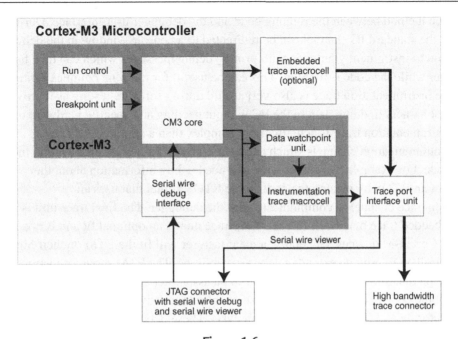

Figure 1.6

The Cortex-M debug architecture is consistent across the Cortex-M family and contains up to three real-time trace units in addition to the run control unit.

The Cortex-M family also has a very advanced debug architecture called CoreSight. The earlier ARM7/9 processors could be debugged through a joint test action group (JTAG) debug interface. This provided a means to download the application code into the on-chip flash memory and then exercise the code with basic run/stop debugging. While a JTAG debugger provided a low-cost way of debugging, it had two major problems. The first was a limited number of breakpoints, generally two with one being required for single stepping code and secondly, when the CPU was executing code the microcontroller became a black box with the debugger having no visibility to the CPU, memory, or peripherals until the microcontroller was halted. The CoreSight debug architecture within the Cortex-M processors is much more sophisticated than the old ARM7 or ARM9 processors. It allows up to eight hardware breakpoints to be placed in code or data regions. CoreSight also provides three separate trace units that support advanced debug features without intruding on the execution of the Cortex CPU. The Cortex-M3 and Cortex-M4 are always fitted with a data watchpoint and trace (DWT) unit and an instrumentation trace macrocell (ITM) unit. The debug interface allows a low-cost debugger to view the contents of memory and peripheral registers "on the fly" without halting the CPU, and the DWT can export a number of watched data, everything that is accessed by the processor, without stealing any cycles from the CPU. The second trace unit is called the instrumentation trace. This trace unit provides a debug communication method between the running code and the debugger user interface. During development, the standard IO channel can be redirected to a console window in the debugger. This allows you to instrument your code with printf() debug messages which can then be read in the debugger while the code is running. This can be useful for trapping complex runtime problems. The instrumentation trace is also very useful during software testing as it provides a way for a test harness to dump data to the PC without needing any specific hardware on the target. The instrumentation trace is actually more complex than a simple UART, as it provides 32 communication channels which can be used by different resources within the application code. For example, we can provide extended debug information about the performance of an RTOS by placing the code in the RTOS kernel that uses an instrumentation trace channel to communicate with the debugger. The final trace unit is called the embedded trace macrocell (ETM). This trace unit is an optional fit and is not present on all Cortex-M devices. Generally, a manufacturer will fit the ETM on their high-end microcontrollers to provide extended debug capabilities. The ETM provides instruction trace information that allows the debugger to build an assembler and High level language trace listing of the code executed. The ETM also enables more advanced tools such as code coverage monitoring and timing performance analysis. These debug features are often a requirement for safety critical and high integrity code development.

Advanced Architectural Features

The Cortex-M3 and Cortex-M4 can also be fitted with another unit to aid high integrity code execution. The memory protection unit allows developers to segment the Cortex-M memory map into regions with different access privileges. We will look at the operating modes of the Cortex-M processor in Chapter 5, but to put it in simple terms, the Cortex CPU can execute the code in a privileged mode or a more restrictive unprivileged mode. The memory protection unit (MPU) can define privileged and unprivileged regions over the 4 GB address space (i.e., code, ram, and peripheral). If the CPU is running in unprivileged mode and it tries to access a privileged region of memory, the MPU will raise an exception and execution will vector to the MPU fault service routine. The MPU provides hardware support for more advanced software designs. For example, you can configure the application code so that an RTOS and low-level device drivers have full privileged access to all the features of the microcontroller while the application code is restricted to its own region of code and data. Like the ETM, the MPU is an optional unit which may be fitted by the manufacturers during design of the microcontroller. The MPU is generally found on high-end devices which have large amounts of flash memory and SRAM. Finally, the Cortex-M3 and Cortex-M4 are interfaced to the rest of the microcontroller through a Harvard bus architecture. This means that they have a port for fetching instructions and constants from code memory and a second port for accessing SRAM and peripherals. We will look at the bus interface more closely in Chapter 5, but in essence, the Harvard bus architecture increases the performance of the Cortex-M processor but does not introduce any additional complexity for the programmer.

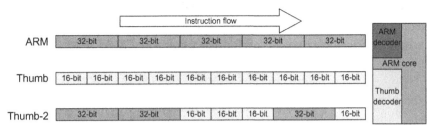

Figure 1.7
Earlier ARM CPUs had two instruction sets, ARM (32 bit) and Thumb (16 bit). The Cortex-M processors have an instruction set called Thumb-2 which is a blend of 16- and 32-bit instructions.

The earlier ARM CPUs, ARM7 and ARM9, supported two instruction sets. This code could be compiled either as 32-bit ARM code or as 16-bit Thumb code. The ARM instruction set would allow code to be written for maximum performance, while Thumb code would achieve a greater code density. During development, the programmer had to decide which function should be compiled with the ARM 32-bit instruction set and which should be built using the Thumb 16-bit instruction set. The linker would then interwork the two instruction sets together. While the Cortex-M processors are code compatible with the original Thumb instruction set, they are designed to execute an extended version of the Thumb instruction set called Thumb-2. Thumb-2 is a blend of 16- and 32-bit instructions that has been designed to be very C friendly and efficient. For even the smallest Cortex-M project, all of the code can be written in a high-level language, typically C, without any need to use an assembler.

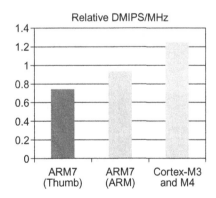

Figure 1.8
The Cortex-M3 and Cortex-M4 Thumb-2 instruction set achieves higher performance levels than either the Thumb or the ARM instruction set running on the ARM7.

With the Thumb-2 instruction set, the Cortex-M3 provides 1.25 DMIPS/MHz, while the ARM7 CPU using the ARM 32-bit instruction set achieves 0.95 DMIPS/MHz, and the Thumb 16-bit instruction set on ARM7 is 0.7 DMIPS/MHz. The Cortex-M CPU has a number of hardware accelerators such as single cycle multiply, hardware division, and bit field manipulation instructions that help boost its performance compared to ARM7-based devices.

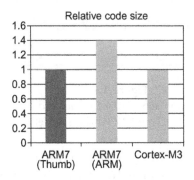

Figure 1.9

The Thumb-2 instruction set of the Cortex-M3 and Cortex-M4 achieves the same code density as the ARM7 Thumb (16 bit) instruction set.

The Thumb-2 instruction set is also able to achieve excellent code density that is comparable to the original 16-bit Thumb instruction set while delivering more processing performance than the ARM 32-bit instruction set.

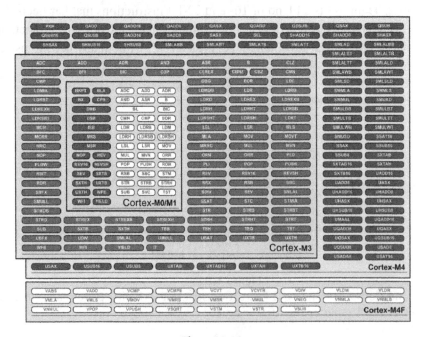

Figure 1.10

The Thumb-2 instruction set scales from 56 instructions on the Cortex-M0 and Cortex-M0+ to up to 169 instructions on the Cortex-M4.

All of the Cortex-M processors use the Thumb-2 instruction set. The Cortex-M0 uses a subset of just 56 instructions and the Cortex-M4 adds the DSP, single instruction multiple data (SIMD), and floating point instructions.

Cortex-M0

The Cortex-M0 was introduced a few years after the Cortex-M3 was released and was in general use. The Cortex-M0 is a much smaller processor than the Cortex-M3 and can be as small as 12 K gates in minimum configuration. The Cortex-M0 is typically designed into microcontrollers that are intended to be very low-cost devices and\or intended for low-power operation. However, the important thing is that once you understand the Cortex-M3, you will have no problem using the Cortex-M0; the differences are mainly transparent to high-level languages.

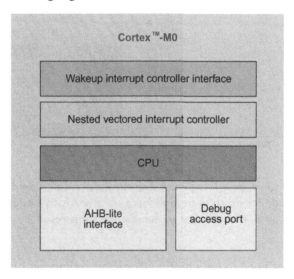

Figure 1.11
The Cortex-M0 is a reduced version of the Cortex-M3 while still keeping the same programmers model.

The Cortex-M0 processor has a CPU that can execute a subset of the Thumb-2 instruction set. Like the Cortex-M3, it has a three-stage pipeline but no branch speculation fetch, therefore branches and jumps within the code will cause the pipeline to flush and refill before execution can resume. The Cortex-M0 also has a von Neumann bus architecture, so there is a single path for code and data. While this makes for a simple design, it can become a bottleneck and reduce performance. Compared to the Cortex-M3, the Cortex-M0 achieves 0.84 DMIPS/MHz, which while less than the Cortex-M3 is still about the same as an ARM7 which has three times the gate count. So, while the Cortex-M0 is at the bottom end of the Cortex-M family, it still packs a lot of processing power. The Cortex-M0 processor has the same NVIC as the Cortex-M3, but it is limited to a maximum of 32 interrupt lines from the microcontroller peripherals. The NVIC also contains the systick timer that is fully compatible with the Cortex-M3. Most RTOS that run on the Cortex-M3 and Cortex-M4 will also run on the Cortex-M0, though the vendor will need to do a dedicated port and recompile the RTOS code. As a developer, the biggest difference you will find between using the Cortex-M0 and the Cortex-M3 is its debug capabilities. While on the Cortex-M3 and Cortex-M4 there is extensive real-time debug support, the Cortex-M0 has a more modest debug architecture. On the Cortex-M0, the DWT unit does not support data trace and the ITM is not fitted, so we are left with basic run control (i.e., run, halt, single stepping and breakpoints, and watchpoints) and on-the-fly memory/peripheral accesses. This is still an enhancement from the JTAG support provided on ARM7 and ARM9 CPUs.

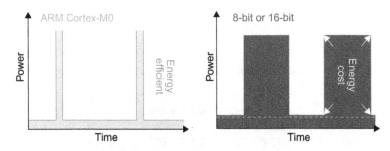

Figure 1.12

The Cortex-M0 is designed to support low-power standby modes. Compared to an 8- or 16-bit MCU, it can stay in sleep mode for much more time because it needs to execute fewer instructions than an 8/16-bit device to achieve the same result.

While the Cortex-M0 is designed to be a high performance microcontroller processor it has a relatively low gate count. This makes it ideal for both low cost and low power devices. The typical power consumption of the Cortex-M0 is 16 μW/MHz when running and almost zero when in its low-power sleep mode. While other 8- and 16-bit architectures can also achieve similar low-power figures, they need to execute far more instructions than the Cortex-M0 to achieve the same end result. This means extra cycles and extra cycles would mean more power consumption. If we pick a good example for Cortex-M0 such as a 16 × 16 multiply, then the Cortex-M0 can perform this calculation in one cycle. In comparison, an 8-bit typical architecture like the 8051 will need at least 48 cycles and a 16-bit architecture will need 8 cycles. This is not only a performance advantage but also an energy efficiency advantage as well.

Table 1.1: Number of Cycles Taken for a 16 × 16 Multiply against Typical 8- and 16-bit Architectures

8-Bit Example (8051)		16-Bit Example	ARM Cortex-M

MOV A, XL ; 2 bytes
MOV B, YL ; 3 bytes
MUL AB; 1 byte
MOV R0, A; 1 byte
MOV R1, B; 3 bytes
MOV A, XL ; 2 bytes
MOV B, YH ; 3 bytes
MUL AB; 1 byte
ADD A, R1; 1 byte
MOV R1, A; 1 byte
MOV A, B ; 2 bytes
ADDC A, #0 ; 2 bytes
MOV R2, A; 1 byte
MOV A, XH ; 2 bytes
MOV B, YL ; 3 bytes
	MUL AB; 1 byte
	ADD A, R1; 1 byte
	MOV R1, A; 1 byte
	MOV A, B ; 2 bytes
	ADDC A, R2 ; 1 bytes
	MOV R2, A; 1 byte
	MOV A, XH ; 2 bytes
	MOV B, YH ; 3 bytes
	MUL AB; 1 byte
	ADD A, R2; 1 byte
	MOV R2, A; 1 byte
	MOV A, B ; 2 bytes
	ADDC A, #0 ; 2 bytes
	MOV R3, A; 1 byte
		MOV R1,&MulOp1
		MOV R2,&MulOp2
		MOV SumLo,R3
		MOV SumHi,R4
		
		(Memory mapped
		  multiply unit)
			MULS r0,r1,r0

Time: 48 clock cycles* **Code size:** 48 bytes		**Time:** 8 clock cycles **Code size:** 8 bytes	**Time:** 1 clock cycle **Code size:** 2 bytes

*cycle count for a single cycle 8051 processor.

Like the Cortex-M3, the Cortex-M0 also has the WIC feature. While the WIC is coupled to the Cortex-M0 processor, it can be placed in a different power domain within the microcontroller.

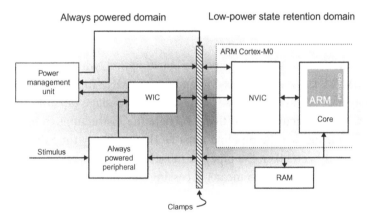

Figure 1.13

The Cortex-M processor is designed to enter low-power modes. The WIC can be placed in a separate power domain.

This allows the microcontroller manufacturer to use their expertise to design very low-power devices where the bulk of the Cortex-M0 processor is placed in a dedicated low-power domain which is isolated from the microcontroller peripheral power domain. These kinds of architected sleep states are critical for designs that are intended to run from batteries.

Cortex-M0+

The Cortex-M0+ processor is the latest generation low-power Cortex-M core. It has complete instruction set compatibility with the Cortex-M0 allowing you to use the same compiler and debug tools. As you might expect, the Cortex-M0+ has some important enhancements over the Cortex-M0.

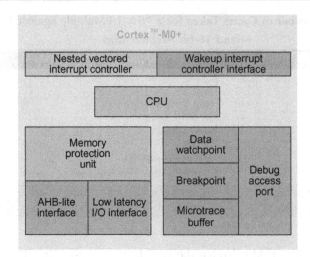

Figure 1.14
The Cortex-M0+ is fully compatible with the Cortex-M0. It has more advanced features such as more processing power and lower power consumption.

The defining feature of the Cortex-M0+ is its power consumption, which is just 11 μW/ MHz compared to 16 μW/MHz for the Cortex-M0 and 32 μW/MHz for the Cortex-M3. One of the Cortex-M0+ key architectural changes is a move to a two-stage pipeline. When the Cortex-M0 and Cortex-M0+ execute a conditional branch, the instructions in the pipeline are no longer valid. This means that the pipeline must be flushed every time there is a branch. Once the branch has been taken the pipeline must be refilled to resume execution. While this impacts on performance it also means accessing the flash memory and each access costs energy as well as time. By moving to a two-stage pipeline, the number of flash memory accesses and hence the runtime energy consumption is also reduced.

Figure 1.15
The Cortex-M0+ has a two-stage pipeline compared to the three-stage pipeline used in other Cortex-M processors.

Another important feature added to the Cortex-M0+ is a new peripheral I/O interface that supports single cycle access to peripheral registers. The single cycle I/O interface is a standard part of the Cortex-M0+ memory map and uses no special instructions or paged addressing. Registers located within the I/O interface can be accessed by normal C pointers from within your application code. The I/O interface allows faster access to peripheral registers with less energy use while being transparent to the application code. The single cycle I/O interface is separate from the advanced high speed bus (AHB) lite external bus interface, so it is possible for the processor to fetch instructions via the AHB lite interface while making a data access to the peripheral registers located within the I/O interface.

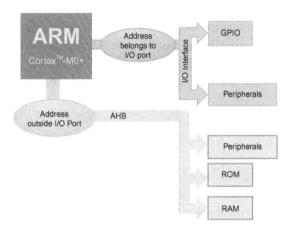

Figure 1.16
The I/O port allows single cycle access to General Purpose IO (GPIO) and peripheral registers.

The Cortex-M0+ is designed to support fetching instructions from 16-bit flash memories. Since most of the Cortex-M0+ instructions are 16 bit, this does not have a major impact on performance but does make the resulting microcontroller design simpler, smaller, and consequently cheaper. The Cortex-M0+ has some Cortex-M3 features missing on the original Cortex-M0. This includes the MPU, which we will look at in Chapter 5, and the ability to relocate the vector table to a different position in memory. These two features provide improved operating system (OS) support and support for more sophisticated software designs with multiple application tasks on a single device.

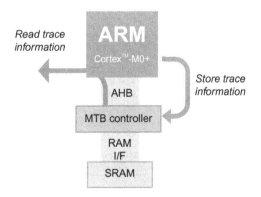

Figure 1.17
The microtrace buffer (MTB) can be configured to record executed instructions into a section of user SRAM. This can be read and displayed as an instruction trace in the PC debugger.

The Cortex-M0+ also has an improved debug architecture compared to the Cortex-M0. As we will see in Chapter 8, it supports the same real-time access to peripheral registers and SRAM as the Cortex-M3 and Cortex-M4. In addition, the Cortex-M0+ has a new debug feature called MTB. The MTB allows executed program instructions to be recorded into a region of SRAM setup by the programmer during development. When the code is halted this instruction trace can be downloaded and displayed in the debugger. This provides a snapshot of code execution immediately before the code was halted. While this is a limited trace buffer, it is extremely useful for tracking down elusive bugs. The MTB can be accessed by standard JTAG/serial wire debug adaptor hardware, for which you do not need an expensive trace tool.

Cortex-M4

While the Cortex-M0 can be thought of as a Cortex-M3 minus some features, the Cortex-M4 is an enhanced version of the Cortex-M3. The additional features on the Cortex-M4 are focused on supporting DSP algorithms. Typical algorithms are transforms such as fast Fourier transform (FFT), digital filters such as finite impulse response (FIR) filters, and control algorithms such as a Proportional Internal Differential (PID) control loop. With its DSP features, the Cortex-M4 has created a new generation of ARM-based devices that can be characterized as digital signal controllers (DSC). These devices allow you to design devices that combine microcontroller type functions with real-time signal processing. In Chapter 7, we will look at the Cortex-M4 DSP extensions in more detail and also how to construct software that combines real-time signal processing with typical event-driven microcontroller code.

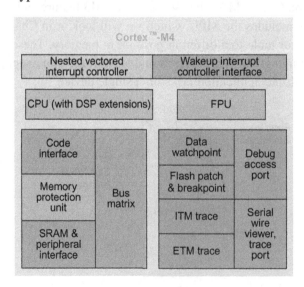

Figure 1.18

The Cortex-M4 is fully compatible with the Cortex-M3 but introduces a hardware floating point unit (FPU) and additional DSP instructions.

The Cortex-M4 has the same basic structure as the Cortex-M3 with the same CPU programmers modes, NVIC, CoreSight debug architecture, MPU, and bus interface. The enhancements over the Cortex-M3 are partly to the instruction set where the Cortex-M4 has additional DSP instructions in the form of SIMD instructions. The hardware multiply accumulate (MAC) has also been improved so that many of the 32×32 arithmetic instructions are single cycle.

Table 1.2: Single Cycle MAC Instructions on the Cortex-M4

Operation	Instructions
16 x 16 = 32	SMULBB, SMULBT, SMULTB, SMULTT
16 x 16 + 32 = 32	SMLABB, SMLABT, SMLATB, SMLATT
16 x 16 + 64 = 64	SMLALBB, SMLALBT, SMLALTB, SMLALTT
16 x 32 = 32	SMULWB, SMULWT
(16 x 32) + 32 = 32	SMLAWB, SMLAWT
(16 x 16) ± (16 x 16) = 32	SMUAD, SMUADX, SMUSD, SMUSDX
(16 x 16) ± (16 x 16) + 32 = 32	SMLAD, SMLADX, SMLSD, SMLSDX
(16 x 16) ± (16 x 16) + 64 = 64	SMLALD, SMLALDX, SMLSLD, SMLSLDX
32 x 32 = 32	MUL
32 ± (32 x 32) = 32	MLA, MLS
32 x 32 = 64	SMULL, UMULL
(32 x 32) + 64 = 64	SMLAL, UMLAL
(32 x 32) + 32 + 32 = 64	UMAAL
32 ± (32 x 32) = 32 (upper)	SMMLA, SMMLAR, SMMLS, SMMLSR
(32 x 32) = 32 (upper)	SMMUL, SMMULR

DSP Instructions

The Cortex-M4 has a set of SIMD instructions aimed at supporting DSP algorithms. These instructions allow a number of parallel arithmetic operations in a single processor cycle.

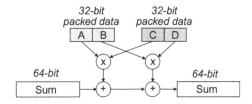

Figure 1.19
The SIMD instructions can perform multiple calculations in a single cycle.

The SIMD instructions work with 16- or 8-bit data which has been packed into 32-bit word quantities. So, for example, we can perform two 16-bit multiplies and sum the result into a 64-bit word. It is also possible to pack the 32-bit works with 8-bit data and perform a quad 8-bit addition or subtraction. As we will see in Chapter 7, the SIMD instructions can be used to vastly enhance the performance of DSP algorithms such as digital filters that are basically performing lots of multiply and sum calculations on a pipeline of data.

The Cortex-M4 processor may also be fitted with a hardware FPU. This choice is made at the design stage by the microcontroller vendor, so like the ETM and MPU you will need to check the microcontroller datasheet to see if it is present. The FPU supports single precision floating point arithmetic calculations using the IEEE 754 standard.

Table 1.3: Cortex-M4 FPU Cycle Times

Operation	Cycle count
Add/Subtract	1
Divide	14
Multiply	1
Multiply accumulate (MAC)	3
Fused MAC	3
Square root	14

On small microcontrollers, floating point math has always been performed by software libraries provided by the compiler tool. Typically, such libraries can take hundreds of instructions to perform a floating point multiply. So, the addition of floating point hardware that can do the same calculation in a single cycle gives an unprecedented performance boost. The FPU can be thought of as a coprocessor that sits alongside the Cortex-M4 CPU. When a calculation is performed, the floating point values are transferred directly from the FPU registers to and from the SRAM memory store, without the need to use the CPU registers. While this may sound involved the entire FPU transaction is managed by the compiler. When you build an application for the Cortex-M4, you can compile code to automatically use the FPU rather than software libraries. Then any floating point calculations in your C code will be carried out on the FPU.

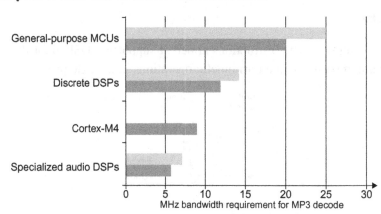

Figure 1.20
MP3 decode benchmark.

With optimized code, the Cortex-M4 can run DSP algorithms far faster than the standard microcontrollers and even some dedicated DSP devices. Of course, the weasel word here is "optimized." This means having a good knowledge of the processor and the DSP algorithm you are implementing and then hand coding the algorithm by making use of compiler intrinsics to get the best level of performance. Fortunately, ARM provides a full open source DSP library that implements many commonly required DSP algorithms as easy to use library functions. We will look at using this library in Chapter 7.

CHAPTER 2

Developing Software for the Cortex-M Family

Introduction

One of the big advantages of using a Cortex-M processor is that it has wide development tool support. There are toolchains available from almost zero cost up to several thousand dollars depending on the depth of your pockets and the type of application you are developing. Today there are five main toolchains that are used for Cortex-M development.

Table 2.1: Cortex Processor Toolchains

1. GNU GCC
2. Greenhills
3. IAR embedded workbench for ARM
4. Keil microcontroller development kit for ARM (MDK-ARM)
5. Tasking VX toolset for ARM

Strictly speaking, the GNU GCC is a compiler linker toolchain and does not include an integrated development environment (IDE) or a debugger. A number of companies have created a toolchain around the GCC compiler by adding their own IDE and debugger to provide a complete development system. Some of these are listed in the appendix; there are quite a few, so this is not a complete list.

Keil Microcontroller Development Kit

The Keil MDK-ARM provides a complete development environment for all Cortex-M-based microcontrollers.

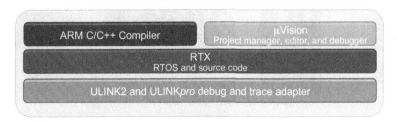

Figure 2.1
The MDK-ARM contains an IDE, compiler, RTOS, and debugger.

The MDK-ARM includes its own development environment called μVision (MicroVision), which acts as an editor, project manager, and debugger. One of the great strengths of the MDK-ARM is that it uses the ARM C compiler. This is a very widely used C\C++ compiler which has been continuously developed by ARM since the first CPUs were created. The MDK-ARM also includes an integrated RTOS called Real Time Executive or RTX. All of the Cortex-M processors are capable of running an OS and we will look at using an RTOS in Chapter 7. As well as including an RTOS, the MDK-ARM also includes a DSP library which can be used on the Cortex-M4 and Cortex-M3 processors. We will look at this library in Chapter 7.

The Tutorial Exercises

There are a couple of key reasons for using the MDK-ARM as the development environment for this book. First, it includes the ARM C compiler that is the industry reference compiler for ARM processors. Second, it includes a software simulator that models the Cortex-M processor and the peripherals of Cortex-M-based microcontrollers. This allows you to get most of the tutorial examples in this book without the need for a hardware debugger or evaluation board. The simulator is a very good way to learn how each Cortex-M processor works as you can get more detailed debug information from the simulator than from a hardware debugger. The MDK-ARM also includes the first RTOS to support the Cortex microcontroller software interface standard (CMSIS) RTOS specification. We will see more of this in Chapter 6, but it is basically a universal application programming interface (API) for Cortex-M RTOS. While we can use the simulator to experiment with the different Cortex-M processors, there comes a point when you will want to run your code on some real hardware. Now there are a number of very low-cost modules that include debugger hardware. The appendix provides URLs to web sites where these boards can be purchased.

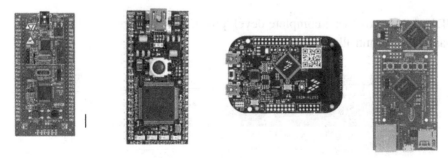

Figure 2.2
Low-cost Cortex-M modules include the STMicroelectronics discovery board, Freescale Freedom board, ARM (NXP) MBED module, and Infineon Technologies Relax board.

The tutorial exercises described in this book are geared toward using the software simulator to demonstrate the different features of the different Cortex-M processors. Matching example sets are also available to run on the different hardware modules where possible. If there are any differences between the instructions for the simulation exercise and the hardware exercise, an updated version of the exercise instructions is included as a PDF in the project directory.

Installation

For the practical exercises in this book, we need to install the MDK-ARM lite toolchain and the example set you plan to use.

First download the Keil MDK-ARM Lite Edition from www.keil.com.

The MDK-ARM Lite Edition is a free version of the full toolchain which allows you to build application code up to 32 K image size. It includes the fully working compiler, RTOS, and simulator and works with compatible hardware debuggers. The direct link for the download is https://www.keil.com/arm/demo/eval/arm.htm.

If the website changes and the link no longer works just the main Keil website at www.keil. com and follow the link to the MDK-ARM tools.

Run the downloaded executable to install the MDK-ARM onto your PC.

Download the example set for the practical exercises.

The different example sets can be found at http://booksite.elsevier.com/9780080982960.

If you do not have any hardware, then download the simulation example set. Otherwise, download the example set for the hardware module you are planning to use.

Once you have downloaded the example set that you are planning to use, run the installer and you are ready to go.

By default, the example set will be installed to C:\ Cortex_M_Tutorial_Exercises.

Exercise Building a First Program

Now that the toolchain and the example sets are installed, we can look at setting up a project for a typical small Cortex-M-based microcontroller. Once the project is set up, we can get familiar with the μVision IDE, build the code, and take our first steps with the debugger.

All of the example sets start by building this project for the simulator, and I recommend that you follow the first exercise described below. If you have a hardware module, a second example in this exercise is used to connect the μVision IDE to the hardware module.

The hardware-based example will have different application code to match the facilities available on the module.

The Blinky Project

In this example, we are going to build a simple project called *blinky*. The code in this project is designed to read a voltage using the microcontroller's ADC. The value obtained from the ADC is then displayed as a bar graph on a small LCD display.

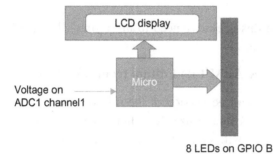

Figure 2.3
The blinky project hardware consists of an analog, voltage source, and external LCD display and a bank of LEDs.

The code also flashes a group of LEDs in sequence. There are eight LEDs attached to port pins on GPIO port B. The speed at which the LEDs flash is set by the ADC value.

Start the µVision IDE by clicking on the UV4 icon.

Once the IDE has launched, close any open project by selecting Project\Close Project from the main menu bar.

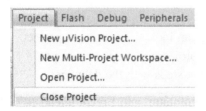

Start a new project by selecting Project\New μVision Project.

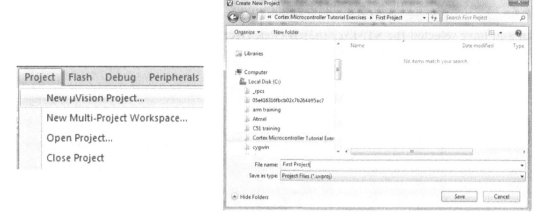

This will open a menu asking for a project name and directory.

You can give the project any name you want but make sure you select the Cortex-M tutorial\first project directory.

This directory contains the C source code files which we will use in our project.

Enter your project name and click save.

Next, select the microcontroller to use in the project.

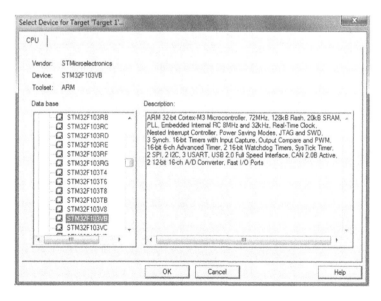

Once you have selected the project directory and saved the project name, a new dialog with a device database will be launched. Here, we must select the microcontroller that we are going to use for this project. Navigate the device database and select STMicroelectronics and the STM32F103RB and click OK. This will configure the project to use the STM32, which includes setting up the correct compiler options, linker script file, simulation model, debugger connection, and flash programming algorithms.

When you have selected the STM32F103RB click OK.

Then, click Yes to add the startup code to the project.

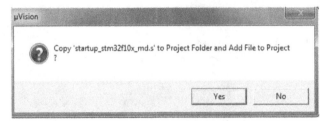

Once you have selected the device, the project wizard will ask if you want to add the STM32 startup code to the project, again say yes to this. The startup file provides the necessary code to initialize the C runtime environment and runs from the processor reset vector to the main() function in your application code. As we will see later, it provides the processor vector table, stack, and variable initialization. It does not provide any hardware initialization for user peripherals or the processor system peripherals. Once you have completed the processor wizard, your project window should have a target folder, and a source group folder. The startup file should be inside the source group folder as shown below.

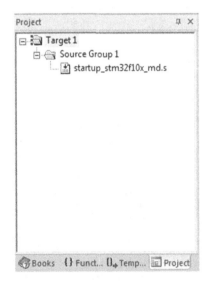

Double click on the startup_stm32F10x.md.s file to open it in the editor.

Click on the Configuration Wizard tab at the bottom of the editor window.

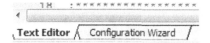

This converts the plain text source file to a view that shows the configuration options within the file.

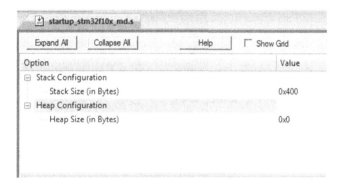

This view is created by XML tags in the source file comments. Changing the values in the configuration wizard modifies the underlying source code. In this case, we can set the size of the stack space and the heap space.

In the project view, click the Books tab at the bottom of the window.

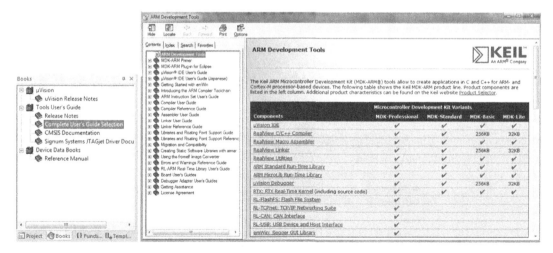

In the books window, the "Complete Users Guide Selection" opens the full help system.

Switch back to the project view and add the project C source files.

Highlight the source group folder in the project window, then right click and select "Add Files to Group Source Group 1."

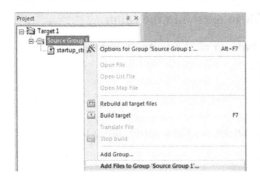

This will open an "Add Files to Group" dialog, which contains the eight ".c" files that are in the first project directory. Select each of these files and add them to the project. It is possible to highlight all the .c files in the add files dialog and add them in one go. Now, the project window should contain all the source files.

Build the project by selecting Project\Build target.

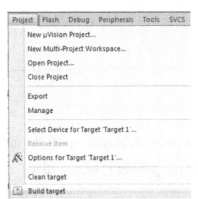

This will compile each of the .c modules in turn and then link them together to make a final application program. The output window shows the result of the build process and reports any errors or warnings.

```
Build Output
Build target 'Target 1'
linking...
Program Size: Code=3684 RO-data=488 RW-data=52 ZI-data=1644
"test.axf" - 0 Error(s), 0 Warning(s).
```

The program size is reported as listed in the following table.

Table 2.2: Linker Data Types

Code	Size of the executable image
RO data	Size of the code constants in the flash memory
RW data	Size of the initialized variables in SRAM
ZI data	Size of the uninitialized variables in SRAM

If errors or warnings are reported in the build window clicking on them will take you to the line of code in the editor window.

Now start the debugger and run the code.

This will connect μVision to the simulation model and download the project image into the simulated memory of the microcontroller. Once the program image has been loaded, the microcontroller is reset and the code is run until it reaches main() ready to start debugging.

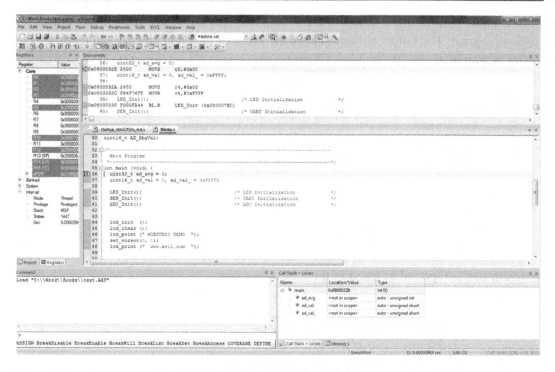

The μVision debugger is divided into a number of windows that allow you to examine and control the execution of your code. The key windows are as follows.

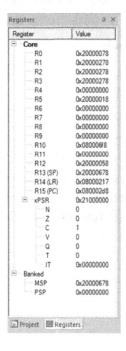

Register Window

The register window displays the current contents of the CPU register file (R0–R15), the program status register (xPSR), and also the main stack pointer (MSP) and the process stack pointer (PSP). We will look at all of these registers in the next chapter.

```
        63:
        64: /*-----------------------------------------------------------------
        65:   MAIN function
        66:   *-------------------------------------------------------------------*/
        67: int main (void) {
0x080002D6 4770      BX              lr
        68:   int32_t num = -1;
0x080002D8 F04F34FF  MOV             r4,#0xFFFFFFFF
        69:   int32_t dir =  1;
0x080002DC 2501      MOVS            r5,#0x01
        70:   uint32_t btns = 0;
        71:
0x080002DE 2600      MOVS            r6,#0x00
        72:   SystemCoreClockUpdate();                        /* Get Core Clock Frequency  */
0x080002E0 F000F944  BL.W            SystemCoreClockUpdate (0x0800056C)
        73:   if (SysTick_Config(SystemCoreClock / 1000)) { /* SysTick 1 msec interrupts */
0x080002E4 482D      LDR             r0,[pc,#180]  ; @0x0800039C
0x080002E6 6800      LDR             r0,[r0,#0x00]
0x080002E8 F44F727A  MOV             r2,#0x3E8
```

Disassembly Window

As its name implies the disassembly window will show you the low level assembler listing interleaved with the high level "C" code listing. One of the great attractions of the Cortex-M family is that all of your project code is written in a high level language such as C\C++. You never need to write low level assembly routines. However, it is useful if you are able to "read" the low level assembly code to see what the compiler is doing. The disassembly window shows the absolute address of the current instruction; next the opcode is shown, which is either a 16- or a 32-bit instruction. The raw opcode is then displayed as an assembler mnemonic. The current location of the program counter is shown by the yellow arrow in the left hand margin. The dark gray blocks indicate the location of executable lines of code.

```
  startup_stm32f10x_md.s    Blinky.c
  30  uint16_t AD_DbgVal;
  31
  32 □/*-----------------------------------------------------------------
  33  |  Main Program
  34  *-------------------------------------------------------------------*/
  35 □int main (void) {
▷ 36    uint32_t ad_avg = 0;
  37    uint16_t ad_val = 0, ad_val_ = 0xFFFF;
  38
▷ 39    LED_Init();                        /* LED Initialization    */
  40    SER_Init();                        /* UART Initialization   */
  41    ADC_Init();                        /* ADC Initialization    */
  42
  43
  44    lcd_init  ();
  45    lcd_clear ();
  46    lcd_print (" MCBSTM32 DEMO  ");
  47    set_cursor(0, 1);
  48    lcd_print ("  www.keil.com  ");
  49
  50
```

The source code window has a similar layout as the disassembly window. This window just displays the high level C source code. The current location of the program counter is shown by the yellow arrow in the left hand margin. The blue arrow shows the location of the cursor. Like the disassembly window, the dark gray blocks indicate the location of executable lines of code. The source window allows you to have a number of project modules open. Each source module can be reached by clicking the tab at the top of the window.

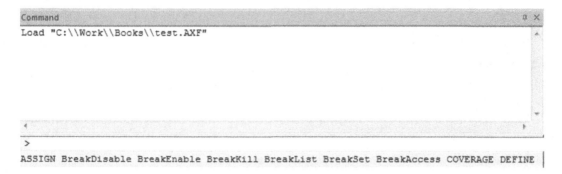

The command window allows you to enter debugger commands to directly configure and control the debugger features. These commands can also be stored in a text file and executed as a script when the debugger starts.

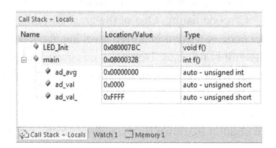

Next to the command window is a group of watch windows. These windows allow you to view local variables, global variables, and the raw memory.

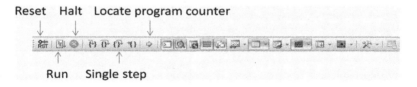

You can control execution of the code through icons on the toolbar. The code can be a single stepped C or assembler line at a time, run at full speed and halted. The same commands are available through the debug menu, which also shows the function key shortcuts that you may prefer.

Start the code running for a few seconds and then press the halt icon.

```
 105
 106    static void delay (int cnt)
 107 ☐ {
▶108        cnt <<= DELAY_2N;
 109
 110        while (cnt--);
 111    }
 112
```

As code is executed in the simulator, the dark gray blocks next to the line number will turn green to indicate that they have been executed. This is a coverage monitor that allows you to verify that your program is executing as expected. In Chapter 8, we will see how to get this information from a real microcontroller.

It is likely that you will halt in a delay function. This is part of the application code that handles the LCD. Currently, the code is waiting for a handshake line to be asserted from the LCD. However, in the simulation only the Cortex-M processor and the microcontroller peripherals are supported. Simulation for external hardware devices like an LCD module has to be created to match the target hardware. This is done through a script file that uses a C-like language to simulate responses from the LCD.

Exit the debugger with the Start/Stop Debug Session option.

Open the Options for Target dialog.

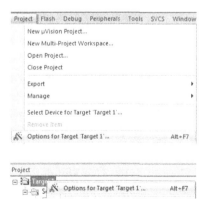

This can be done in the project menu by right clicking the project name and selecting Options for Target or by selecting the same option in the project menu from the main toolbar.

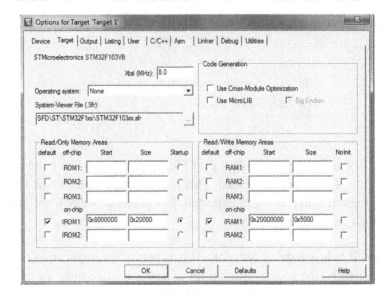

The Options for Target dialog holds all of the global project settings.

Now select the Debug tab.

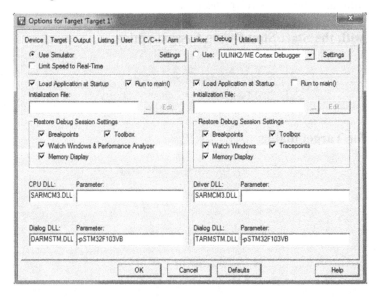

The Debug menu is split into two halves. The simulator options are on the left and the hardware debugger is on the right. Currently, the Use Simulator option is set.

Add the LCD simulation script.

File explorer

Press the file explorer button and add the Dbg_sim.ini file, which is in the first project directory as the debugger initialization file.

The script file uses a C-like language to model the external hardware. All of the simulated microcontroller "pins" appear as virtual registers that can be read from and written to by the script. The debug script also generates a simulated voltage for the ADC. The script for this is shown below. This generates a signal that ramps up and down and it is applied to the virtual register ADC1_IN1, which is channel 1 of ADC convertor 1. The twatch function reads the simulated clock of the processor and halts the script for a specified number of cycles.

```
Signal void Analog (float limit) {
   float volts;
   printf ("Analog (%f) entered.\n", limit);
   while (1) {         /* forever */
```

```
   volts = 0;
   while (volts <= limit) {
     ADC1_IN1 = volts;        /* analog input-2 */
     twatch (250000);          /* 250000 Cycles Time-Break */
     volts + = 0.1;          /* increase voltage */
   }
   volts = limit;
   while (volts >= 0.0) {
     ADC1_IN1 = volts;
     twatch (250000);          /* 250000 Cycles Time-Break */
     volts - = 0.1;          /* decrease voltage */
   }
  }
 }
```

Click OK to close the Options for Target dialog.

Start the debugger.

Set a breakpoint on the main while loop in blinky.c.

```
50
51    SysTick_Config(SystemCoreClock / 100);
52
53    while (1) {
54
55        /* AD converter input
56        if (AD_done) {
57            AD_done = 0;
```

You can set a breakpoint by moving the mouse cursor into a dark gray block next to the line number and left clicking. A breakpoint is marked by a red dot.

Start executing the code.

With the simulation script in place we will be able to execute all of the LCD code and reach the breakpoint.

```
51    SysTick_Config(SystemCoreClock / 100);
52
53    while (1) {
54
55        /* AD converter input
56        if (AD_done) {
57            AD_done = 0;
```

Now spend a few minutes exploring the debugger run control.

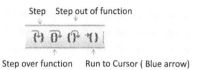

Use the single step commands, set a breakpoint, and start the simulator running at full speed. If you lose what is going on, exit the debugger by selecting debug/start/stop the debugger and then restart again.

Add a variable to the watch window.

Once you have finished familiarizing yourself with the run control commands within the debugger, locate the main() function within blinky.c. Just above main() is the declaration for a variable called AD_DbgVal. Highlight this variable, right click, and select Add AD_DbgVal to Watch 1.

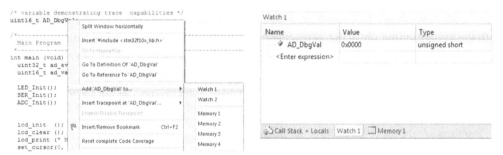

Now start running the code and you will be able to see the ADC variable updating in the watch window.

The simulation script is feeding a voltage to the simulated microcontroller ADC which in turn provides converted results to the application code.

Now add the same variable to the logic trace window.

The μVision debugger also has a logic trace feature that allows you to visualize the historical values of a given global variable. In the watch window (or the source code window), highlight the AD_DbgVal variable name, right click, and select Add AD_DbgVal to Logic Analyzer.

If the logic analyzer window does not open automatically select it from the toolbar.

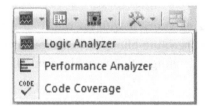

Now with the code running press the autoscale button, which will set the minimum and maximum values, and also click the zoom out button to get to a reasonable time scale. Once this is done, you will be able to view a trace of the values stored in the AD_DbgVal variable. You can add any other global variables or peripheral registers to the logic analyzer window.

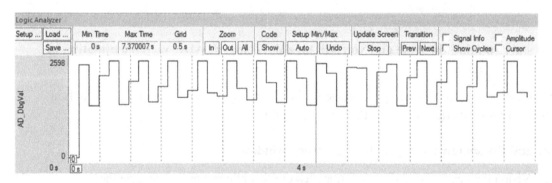

Now view the state of the user peripherals.

The simulator has a model of the whole microcontroller not just the Cortex-M processor, so it is possible to examine the state of the microcontroller peripherals directly.

Select Peripherals\General Purpose IO\GPIOB.

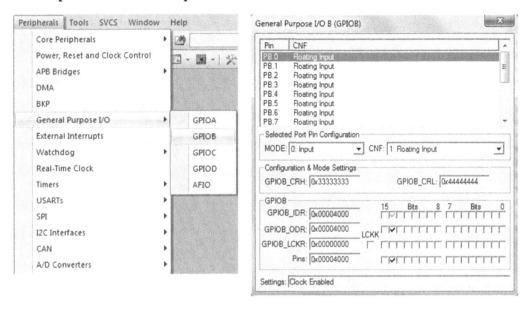

This will open a window that displays the current state of the microcontroller IO Port B. As the simulation runs, we can see the state of the port pins. If the pins are configured as inputs, we can manually set and clear them by clicking the individual Pins boxes.

You can do the same for the ADC by selecting ADC1.

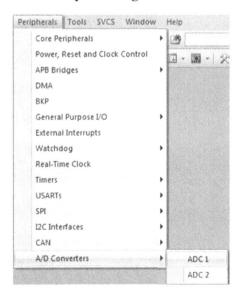

When the code is running it is possible to see the current configuration of the ADC and the conversion results. You can also manually set the input voltage by entering a fresh value in the Analog Inputs boxes.

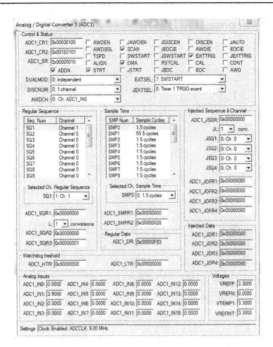

The simulator also provides a terminal that provides an I\O channel for the microcontroller's UARTS.

Select View\Serial Windows\UART #1.

This opens a console-type window that displays the output from a selected UART and also allows you to input values.

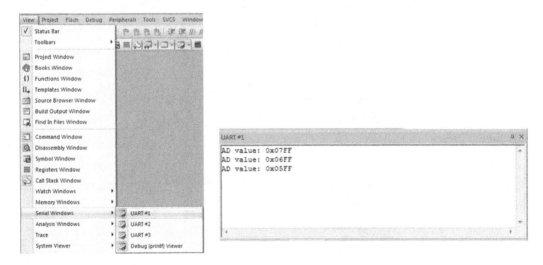

The simulator also boasts some advanced analysis tools including trace, code coverage, and performance analysis.

Open the View\Trace menu and select Trace Data and Enable Trace Recording.

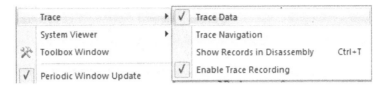

This will open the instruction trace window. This records a history of each instruction executed.

Instruction trace is extremely valuable when tracking down complex bugs and validating software. In Chapter 8, we will see how to get instruction trace from a real microcontroller.

Now open the View/Analysis Windows/Code Coverage and View/Analysis Windows/Performance Analyzer windows.

The performance analysis window shows the number of calls to a function and its cumulative runtime.

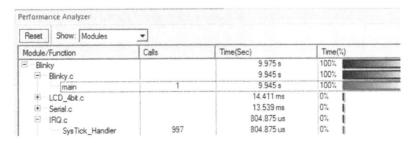

The code coverage window provides a digest of the number of lines executed and partially executed in each function.

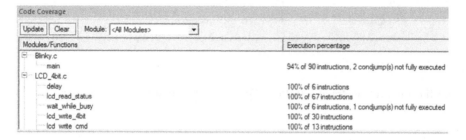

Both code coverage and performance analysis are essential for validating and testing software. In Chapter 8, we will see how this information can be obtained from a real microcontroller.

Project Configuration

Now that you are familiar with the basic features of the debugger, we can look in more detail at how the project code is constructed.

First quit the debugger by selecting Debug\Start/Stop Debug Session.

Open the Options for Target dialog.

All of the key project settings can be found in the Options for Target dialog.

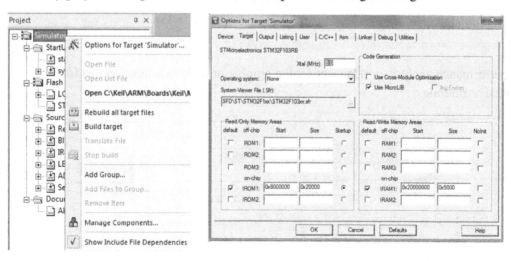

The target tab defines the memory layout of the project. A basic template is defined when you create the project. On this microcontroller, there are 128 K of internal flash memory and 20 K of SRAM. If you need to define a more complex memory layout, it is possible to create additional memory regions to subdivide the volatile and nonvolatile memories.

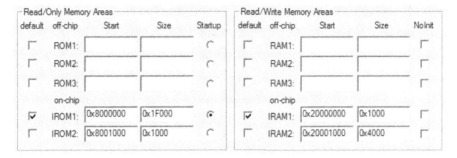

The more complex memory map above has split the internal flash into two blocks and defined the lower flash block as the default region for code and constant data. Nothing will be placed into the upper block unless you explicitly tell the linker to do this. Similarly, the SRAM has been split into two regions and the upper region is unused unless you explicitly tell the linker to use it. When the linker builds the project, it looks for the "RESET" code label. The linker then places the reset code at the base of the code region designated as the startup region. The initial startup code will write all of the internal SRAM to zero unless you tick the NoInit box for a given SRAM region. Then the SRAM will be left with its startup garbage values. This may be useful if you want to allow for a soft reset where some system data is maintained.

If you want to place objects (code or data) into an unused memory region, select a project module, right click, and open its local options.

In its local options, the Memory Assignment boxes allow you to force the different memory objects in a module into a specific code region.

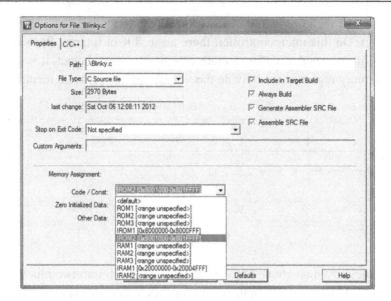

Back in the main Options for Target menu, there is an option to set the external crystal frequency used by the microcontroller.

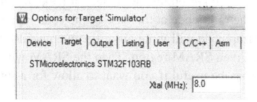

Often this will be a standard value that can be multiplied by the internal phase locked loop oscillator of the microcontroller to reach the maximum clock speed supported by the microcontroller. This option is only used to provide the input frequency for the simulation model and nothing else.

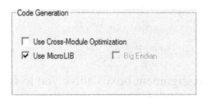

The Keil MDK-ARM comes with two ANSI library sets. The first is the standard library that comes with the ARM compiler. This is fully compliant with the current ANSI standard and as such has a large code footprint for microcontroller use. The second library set is the Keil MicroLIB; this library has been written to an earlier ANSI standard, that is, the C99 standard. This version of the ANSI standard is more in tune with the needs of microcontroller users.

Table 2.3: Size Comparison between the Standard ARM ISO Libraries and the Keil Microlibrary

Processor	Object		Standard	MicroLib	% Savings
Cortex-M0(+)	Thumb	Library total	16,452	5996	64%
		RO total	19,472	9016	54%
Cortex-M3\M4	Thumb-2	Library total	15,018	5796	63%
		RO total	18,616	8976	54%

By selecting MicroLIB you will save at least 50% of the ANSI library code footprint verses the ARM compiler libraries. So try to use MicroLIB wherever possible. However, it does have some limitations, most notably it does not support all of the functions in standard library and double precision floating point calculations. In a lot of applications, you can live with this. Since most small microcontroller based embedded applications do not use these features you can generally use microLIB as your default library.

The Use Cross-Module Optimization tick box enables two pass linking process that fully optimizes your code. When you use this option, the code generation is changed and the code execution may no longer map directly to the C source code. So do not use this option when you are testing and debugging code as you will not be able to accurately follow it within the debugger source code window. We will look at the other options, system viewer file, and operating system in later chapters.

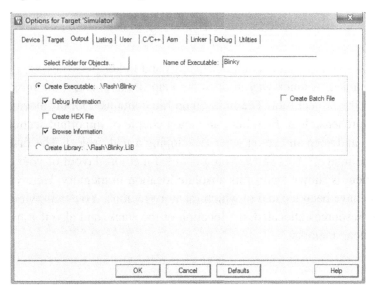

The Output menu allows you to control the final image of your project. Here, we can choose between generating a standalone program by selecting Create Executable and we can create a library that can be added to another project. The default is to create a standalone project with debug information. When you get toward the end of a project, you will need to select the Create HEX File option to generate a HEX32 file, which can be used with a production programmer. If you want to build the project outside of μVision, select Create Batch File and this will produce a <Project name>.bat DOS batch file that can be run from another program to rebuild the project outside of the IDE. By default, the name of the final image is always the same as your project name. If you want to change this, simply change the Name of Executable field. You can also select a directory to store all of the project, compiler, and linker generated files. This ensures that the original project directory only contains your original source code. This can make life easier when you are archiving projects.

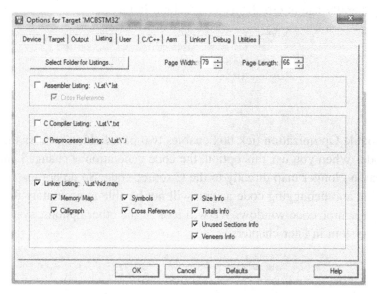

The Listing tab allows you to enable compiler listings and linker map files. By default, the linker map file is enabled. A quick way to open the map file is to select the project window and double click on the project root. The linker map file contains a lot of information, which seems incomprehensible at first, but there are a couple of important sections that you should learn to read and keep an eye on when developing a real project. The first is the "memory map of the image." This shows you a detailed memory layout of your project. Each memory segment is shown against its absolute location in memory. Here you can track which objects have been located to which memory regions. You can review the total amount of memory resources allocated, the location of the stack, and also if it has enabled the location of the heap memory.

```
Image Entry point : 0x000000cd

Load Region LR_IROM1 (Base: 0x00000000, Size: 0x00000d98, Max: 0x00040000, ABSOLUTE)

  Execution Region ER_IROM1 (Base: 0x00000000, Size: 0x00000d78, Max: 0x00040000, ABSOLUTE)

  Base Addr    Size          Type    Attr    Idx    E Section Name          Object

  0x00000000   0x000000cc    Data    RO        3      RESET                 startup_lpc17xx.o
  0x000000cc   0x00000000    Code    RO      267    * .ARM.Collect$$$$00000000  mc_w.l(entry.o)
  0x000000cc   0x00000004    Code    RO      533      .ARM.Collect$$$$00000001  mc_w.l(entry2.o)
  0x000000d0   0x00000004    Code    RO      536      .ARM.Collect$$$$00000004  mc_w.l(entry5.o)
  0x000000d4   0x00000000    Code    RO      538      .ARM.Collect$$$$00000008  mc_w.l(entry7b.o)
  0x000000d4   0x00000008    Code    RO      539      .ARM.Collect$$$$00000009  mc_w.l(entry8.o)
  0x000000dc   0x00000004    Code    RO      534      .ARM.Collect$$$$00002712  mc_w.l(entry2.o)
  0x000000e0   0x00000024    Code    RO        5      .text                 startup_lpc17xx.o
  0x00000104   0x00000030    Code    RO      100      .text                 retarget.o
  0x00000134   0x00000080    Code    RO      127      .text                 serial.o
  0x000001b4   0x000000ac    Code    RO      154      .text                 led.o
  0x00000260   0x00000074    Code    RO      180      .text                 irq.o
  0x000002d4   0x0000001e    Code    RO      542      .text                 mc_w.l(llshl.o)
  0x000002f2   0x00000002    Code    RO      566      i.__scatterload_null  mc_w.l(handlers.o)
  0x000002f4   0x00000008    PAD
  0x000002fc   0x00000004    Code    RO        4      .ARM.__at_0x02FC      startup_lpc17xx.o
  0x00000300   0x00000310    Code    RO       15      .text                 system_lpc17xx.o
  0x00000610   0x000000d8    Code    RO      207      .text                 blinky.o
  0x000006e8   0x000000c4    Code    RO      239      .text                 adc.o
  0x000007ac   0x00000062    Code    RO      270      .text                 mc_w.l(uldiv.o)
  0x0000080e   0x00000020    Code    RO      544      .text                 mc_w.l(llushr.o)
  0x0000082e   0x00000002    PAD
  0x00000830   0x00000024    Code    RO      557      .text                 mc_w.l(init.o)
  0x00000854   0x00000020    Code    RO      479      i.__0printf$8         mc_w.l(printf8.o)
  0x00000874   0x00000028    Code    RO      481      i.__0sprintf$8        mc_w.l(printf8.o)
  0x0000089c   0x0000000e    Code    RO      565      i.__scatterload_copy  mc_w.l(handlers.o)
  0x000008aa   0x0000000e    Code    RO      567      i.__scatterload_zeroinit  mc_w.l(handlers.o)
  0x000008b8   0x00000420    Code    RO      486      i._printf_core        mc_w.l(printf8.o)
  0x00000cd8   0x00000026    Code    RO      487      i._printf_post_padding  mc_w.l(printf8.o)
  0x00000cfe   0x00000030    Code    RO      488      i._printf_pre_padding  mc_w.l(printf8.o)
  0x00000d2e   0x0000000a    Code    RO      490      i._sputc              mc_w.l(printf8.o)
  0x00000d38   0x00000020    Data    RO      155      .constdata            led.o
  0x00000d58   0x00000020    Data    RO      563      Region$$Table         anon$$obj.o

  Execution Region RW_IRAM1 (Base: 0x10000000, Size: 0x00000000, Max: 0x00004000, ABSOLUTE)

  **** No section assigned to this execution region ****

  Execution Region RW_IRAM2 (Base: 0x2007c000, Size: 0x00000230, Max: 0x00008000, ABSOLUTE)

  Base Addr    Size          Type    Attr    Idx    E Section Name          Object

  0x2007c000   0x00000004    Data    RW       16      .data                 system_lpc17xx.o
  0x2007c004   0x00000008    Data    RW      101      .data                 retarget.o
  0x2007c00c   0x0000000d    Data    RW      181      .data                 irq.o
  0x2007c019   0x00000001    PAD
  0x2007c01a   0x00000004    Data    RW      240      .data                 adc.o
  0x2007c01e   0x00000002    PAD
  0x2007c020   0x0000000a    Zero    RW      208      .bss                  blinky.o
  0x2007c02a   0x00000006    PAD
  0x2007c030   0x00000200    Zero    RW        1      STACK                 startup_lpc17xx.o
```

The second section gives you a digest of the memory resources required by each module and library in the project together with details of the overall memory requirements. The image memory usage is broken down into the code data size. The code data size is the amount of nonvolatile memory used to store the initializing values to be loaded into RAM variables on startup. In simple projects, this initializing data is held as a simple ROM table which is written into the correct RAM locations by the startup code. However, in projects with large amounts of initialized data, the compiler will switch strategies and use a compression algorithm to minimize the size of the initializing data. On startup, this table is decompressed before the data is written to the variable locations in memory. The RO data entry lists the amount of nonvolatile memory used to store code literals. The SRAM usage is split into initialized RW data and uninitialized ZI data.

Image component sizes

Code (inc. data)	RO Data	RW Data	ZI Data	Debug	Object Name
196	30	0	4	0	16831	adc.o
216	50	0	0	10	16612	blinky.o
116	20	0	13	0	641	irq.o
172	16	32	0	0	1317	led.o
48	0	0	8	0	2658	retarget.o
128	18	0	0	0	17327	serial.o
40	12	204	0	512	880	startup_lpc17xx.o
784	76	0	4	0	5105	system_lpc17xx.o

Code (inc. data)	RO Data	RW Data	ZI Data	Debug	
1700	222	268	32	528	61371	Object Totals
0	0	32	0	0	0	(incl. Generated)
0	0	0	3	6	0	(incl. Padding)

Code (inc. data)	RO Data	RW Data	ZI Data	Debug	Library Member Name
0	0	0	0	0	0	entry.o
8	4	0	0	0	0	entry2.o
4	0	0	0	0	0	entry5.o
0	0	0	0	0	0	entry7b.o
8	4	0	0	0	0	entry8.o
30	0	0	0	0	0	handlers.o
36	8	0	0	0	68	init.o
30	0	0	0	0	68	llshl.o
32	0	0	0	0	68	llushr.o
1224	62	0	0	0	504	printf8.o
98	0	0	0	0	92	uldiv.o

Code (inc. data)	RO Data	RW Data	ZI Data	Debug	
1480	78	0	0	0	800	Library Totals
10	0	0	0	0	0	(incl. Padding)

Code (inc. data)	RO Data	RW Data	ZI Data	Debug	Library Name
1470	78	0	0	0	800	mc_w.l

Code (inc. data)	RO Data	RW Data	ZI Data	Debug	
1480	78	0	0	0	800	Library Totals

Code (inc. data)	RO Data	RW Data	ZI Data	Debug	
3180	300	268	32	528	61027	Grand Totals
3180	300	268	32	528	61027	ELF Image Totals
3180	300	268	32	0	0	ROM Totals

```
Total RO  Size (Code + RO Data)              3448 (  3.37kB)
Total RW  Size (RW Data + ZI Data)            560 (  0.55kB)
Total ROM Size (Code + RO Data + RW Data)    3480 (  3.40kB)
```

The next tab is the User tab. This allows you to add external utilities to the build process. The menu allows you to run a utility program to pre- or postprocess files in the project. A utility program can also be run before each module is compiled. Optionally, you can also start the debugger once the build process has finished.

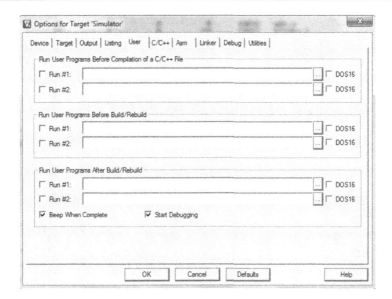

The code generated by the compiler is controlled by the C/C++ tab. This controls the code generation for the whole project. However, the same menu is available in the local options for each source module. This allows you to have global build options and then different build options for selected modules. In the local options menu, the option tick boxes are a bit unusual in that they have three states.

They can be unchecked, checked with a solid black tick, or checked with a gray tick. Here, the gray tick means "inherit the global options" and this is the default state.

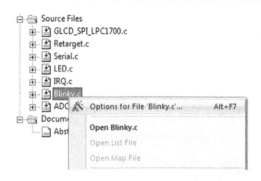

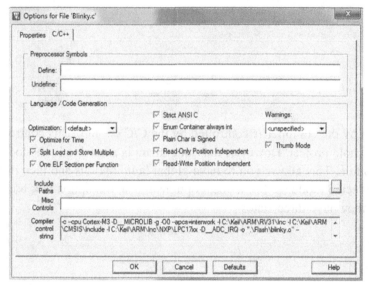

The most important option in this menu is the optimization control. During development and debugging you should leave the optimization level at zero. Then, the generated code maps directly to the high level "C" source code and it is easy to debug. As you increase the optimization level, the compiler will use more and more aggressive techniques to optimize the code. At the high optimization level, the generated code no longer maps closely to the original source code, which then makes using the debugger very difficult. For example, when you single step the code, its execution will no longer follow the expected path through the source code. Setting a breakpoint can also be hit and miss as the generated code may not exist on the same line as the source code. By default, the compiler will generate the smallest image. If you need to get the maximum performance, you can select the Optimize for Time option. Then, the compiler strategy will be changed to generate the fastest executable code.

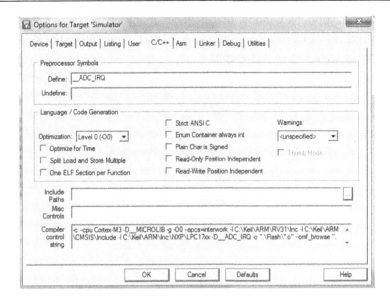

The compiler menu also allows you to enter any #defines that you want to pass to the source module when it is being compiled. If you have structured your project over several directories, you may also add local include paths to directories with project header files. The Misc Controls text box allows you to add any compiler switches that are not directly supported in the main menu. Finally, the full compiler control string is displayed. This includes the CPU options, project include paths and library paths, and the make dependency files.

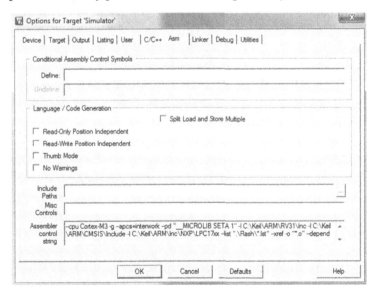

There is also an assembler options window that includes many of the same options as the
C/C++ menu. However, most Cortex-M projects are written completely in C/C++, so
with luck you will never have to use this menu!

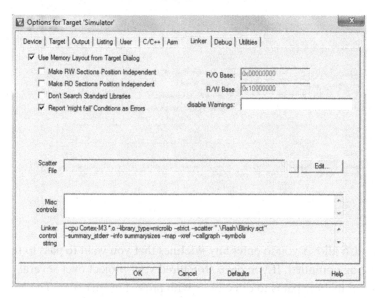

By default, the Linker menu imports the memory layout from the Target menu. This
memory layout is converted into a linker "scatter" file. The scatter file provides a text
description of the memory layout to the linker so it can create a final image. An example of
a scatter file is shown below.

```
*************************************************************
; *** Scatter-Loading Description File generated by µVision ***
; *************************************************************
LR_IROM1 0×00000000 0×00040000 {   ; load region size_region
  ER_IROM1 0×00000000 0×00040000 {   ; load address = execution address
    *.o (RESET, +First)
    *(InRoot$$Sections)
    .ANY (+RO)
  }
  RW_IRAM1 0×10000000 0×00004000 {   ; RW data
    .ANY (+RW +ZI)
  }
  RW_IRAM2 0×2007C000 0×00008000 {
    .ANY (+RW +ZI)
  }
}
```

The scatter file defines the ROM and RAM regions and the program segments that need to be placed in each segment. In the above example, the scatter file first defines a ROM region of 256 K. All of this memory is allocated in one bank. The scatter file also tells the linker to place the reset segment containing the vector table at the start of this section. Next, the scatter file then tells the linker to place all the remaining nonvolatile segments in this region. The scatter file then defines two banks of RAM of 16 and 32 K. The linker is then allowed to use both pages of RAM for initialized and uninitialized variables. This is a simple memory layout that maps directly onto the microcontroller's memory. If you need to use a more sophisticated memory layout, you can add extra memory regions in the Target menu and this will be reflected in the scatter file. If, however, you need a complex memory map which cannot be defined through the Target menu, then you will need to write your own scatter file. The trick here is to get as close as you can with the Target menu and then hand edit the scatter file.

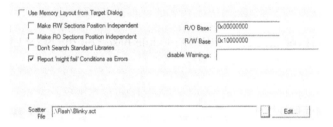

If you are using your own scatter file, you must then uncheck the Use Memory Layout from Target Dialog box and then manualy select the new scatter file using the Scatter File text box.

Hardware Debug

If you have downloaded an example set for a specific hardware module, the first project directory will contain a subdirectory named after the module you are using.

In μVision, select Project\Open Project.

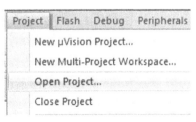

Open the project located in the hardware module directory.

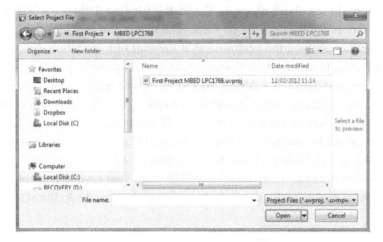

The project shown here is for the ARM MBED module based on the NXP LPC1768.

Build the project.

The project is created and builds in exactly the same way as the simulator version.

Open the Options for Target dialog and the Debug tab.

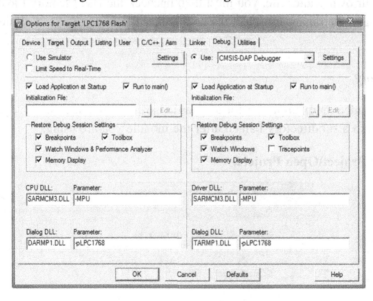

In the Debug menu, the Use option has been switched to select a hardware debugger rather than the simulator.

Now open the Utilities menu.

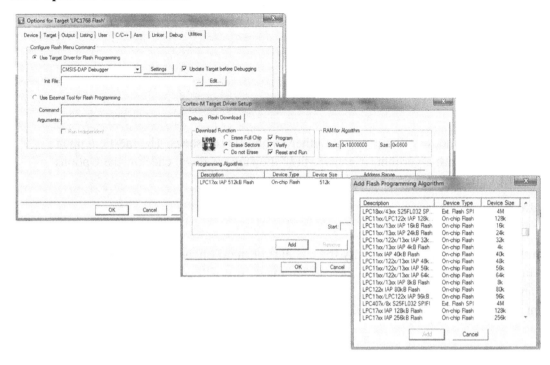

The Utilities menu allows you to select a tool to program the microcontroller flash memory. This will normally be the same as the debugger interface selected in the Debug menu. Pressing the setting button allows you to add the flash algorithm for the microcontroller.

The most common flash programming problems are listed below.

One point worth noting in the Utility menu is the Update Target before Debugging tick box.

When this option is ticked, the flash memory will be reprogrammed when the debugger starts. If it is not checked, then you must manually reprogram the flash by selecting Flash\Download from the main toolbar.

If there is a problem programming the flash memory, you will get the following error window pop-up.

The build output window will report any further diagnostic messages. The most common error is a missing flash algorithm. If you see the following message, check if the Options for Target\Utilities menu is configured correctly.

```
No Algorithm found for: 00000000H - 000032A3H
Erase skipped!
```

When the debugger starts, it will verify the contents of the flash against an image of the program. If the flash does not match the current image, you will get a memory mismatch error and the debugger will not start. This means that the flash image is out of date and the current version needs to be downloaded into the flash memory.

Select Cancel to close both of these dialogs without making any changes.

Start the debugger.

When the debugger starts, it is now connected to the hardware and will download the code into the flash memory of the microcontroller and allow the debugger interface to control the real microcontroller in place of the simulation model.

Experiment with the debugger interface now that it is connected to the real hardware.

You will note that some of the features available in the simulator are not present when using the hardware module. These are the instruction trace, code coverage, and performance analysis windows. These features are available with hardware debug, but you need a more advanced hardware interface to get them.

Cortex-M Architecture

Introduction

In this chapter, we will take a closer look at the Cortex-M processor architecture. The bulk of this chapter will concentrate on the Cortex-M3 processor. Once we have a firm understanding of the Cortex-M3, we will look at the key differences in the Cortex-M0, M0+, and M4. There are a number of exercises throughout the chapter. These exercises will give you a deeper understanding of each topic and can be used as a reference when developing your own code.

Cortex-M Instruction Set

As we described in Chapter 1, the Cortex-M processors are Reduced Instruction Set Computer (RISC)-based processors and as such have a small instruction set. The Cortex-M0 has just 56 instructions, the Cortex-M3 has 74, and the Cortex-M4 has 137 with an option of additional 32 instructions for the FPU. The ARM CPUs, ARM7 and ARM9, which were originally used in microcontrollers, have two instruction sets: the ARM (32 bit) instruction set and the Thumb (16 bit) instruction set. The ARM instruction set was designed to get maximum performance from the CPU while the Thumb instruction set featured good code density to allow programs to fit into the limited memory resources of a small microcontroller. The developer had to decide which function was compiled with the ARM instruction set and which was compiled with the Thumb instruction set. Then the two groups of functions could be "interworked" together to build the final program. The Cortex-M instruction set is based on the earlier 16-bit Thumb instruction set found in the ARM processors but extends that set to create a combined instruction set with a blend of 16- and 32-bit instructions.

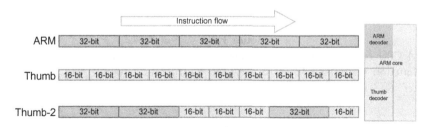

Figure 3.1

The ARM7 and ARM9 CPU had separate 32- and 16-bit instruction sets. The Cortex-M processor has a single instruction set that is a blend of 16- and 32-bit instructions.

The Designer's Guide to the Cortex-M Processor Family.

The Cortex-M instruction set, called Thumb-2, is designed to make this much simpler and more efficient. The good news is that your whole Cortex-M project can be coded in a high level language such as C\C++ without the need for any hand-coded assembler. It is useful to be able to "read" Thumb-2 assembly code via a debugger disassembly window to check what the compiler is up to, but you will never need to write an assembly routine. There are some useful Thumb-2 instructions that are not reachable using the C language but most compiler toolchains provide intrinsic instructions which can be used to access these instructions from within your C code.

Programmer's Model and CPU Registers

The Cortex-M processors inherit the ARM RISC's load and store method of operation. This means that to do any kind of data processing, instruction such as ADD and SUBTRACT the data must first be loaded into the CPU registers; the data processing instruction is then executed and the result is stored back in the main memory. This means that code executing on a Cortex-M processor revolves around the central CPU registers.

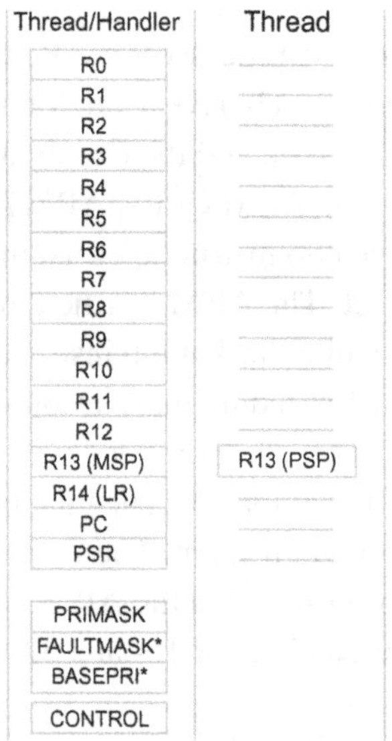

Thread/Handler	Thread
R0	
R1	
R2	
R3	
R4	
R5	
R6	
R7	
R8	
R9	
R10	
R11	
R12	
R13 (MSP)	R13 (PSP)
R14 (LR)	
PC	
PSR	
PRIMASK	
FAULTMASK*	
BASEPRI*	
CONTROL	

Figure 3.2
The Cortex-M CPU registers consist of 16 data registers, a program status register, and four special function registers. R13–R15 are used as the stack pointer, link register, and program counter. R13 is a banked register which allows the Cortex-M CPU to operate with dual stacks.

On all Cortex-M processors, the CPU register file consists of 16 data registers followed by the program status register and a group of configuration registers. All of the data registers (R0–R15) are 32 bits wide and can be accessed by all of the Thumb-2 load and store instructions. The remaining CPU registers may only be accessed by two dedicated instructions, move general register to special register (MRS) and move special register to general register (MSR), where the "special registers" are PRIMASK, FAULTMASK, BASEPRI, and CONTROL. The registers R0–R12 are general user registers and are used by the compiler as it sees fit. The registers R13–R15 have special functions. R13 is used by the compiler as the stack pointer; this is actually a banked register with two R13 registers. When the Cortex-M processor comes out of reset, this second R13 register is not enabled and the processor runs in a "simple" mode with one stack pointer. It is possible to enable the second R13 register by writing to the Cortex control register. The processor will then be configured to run with two stacks. We will look at this in more detail in Chapter 5 "Advanced Architecture Features" but for now we will use the Cortex-M processor in its default mode. After the stack pointer we have R14, the link register. When a procedure is called the return address is automatically stored in R14. Since the Thumb-2 instruction set does not contain a RETURN instruction when the processor reaches the end of a procedure it uses the branch instruction on R14 to return. Finally, R15 is the program counter. You can operate on this register just like all the others but you will not need to do this during normal program execution. The CPU registers PRIMASK, FAULTMASK, and BASEPRI are used to temporarily disable interrupt handling and we will look at these later in this chapter.

Program Status Register

The PSR as its name implies contains all the CPU status flags.

Figure 3.3
The PSR contains several groups of CPU flags. These include the condition codes (NZCVQ), interrupt continuable instruction (ICI) status bits, If Then (IT) flag, and current exception number.

The PSR has a number of alias fields that are masked versions of the full register. The three alias registers are the application program status register (APSR), interrupt program status register (IPSR), and the execution program status register (EPSR). Each of these alias registers contains a subset of the full register flags and can be used as a shortcut if you need to access part of the PSR. The PSR is generally referred to as the xPSR to indicate the full register rather than any of the alias subsets.

Figure 3.4
The PSR has three alias registers that provide access to specific subregions of the PSR. Hence, the generic name for the PSR is xPSR.

In a normal application program, your code will not make explicit access to the xPSR or any of its alias registers. Any use of the xPSR will be made by compiler generated code. As a programmer you need to have an awareness of the xPSR and the flags contained in it.

The most significant four bits of the xPSR are the condition code bits—Negative, Zero, Carry, and oVerflow. These will be set and cleared depending on the results of a data processing instruction. The result of Thumb-2 data processing instructions can set or clear these flags. However, updating these flags is optional.

```
SUB R8, R6, #240   Perform a subtraction and do not update the condition code flags
SUBS R8, R6, #240   Perform a subtraction and update the condition code flags
```

This allows the compiler to perform an instruction that updates the condition code flags, then perform some additional instructions that do not modify the flags and then perform a conditional branch on the state of the xPSR condition codes. Following the four condition code flags is a further instruction flag, the Q bit.

Q Bit and Saturated Math Instructions

The Q bit is the saturation flag. The Cortex-M3 and Cortex-M4 processors have a special set of instructions called the saturated math instructions. If a normal variable reaches its maximum value and you increment it further, it will roll round to zero. Similarly, if a variable reaches its minimum value and is then decremented, it will roll round to the maximum value.

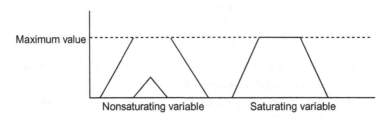

Figure 3.5
A normal variable will rollover to zero when it hits its maximum value. This is very dangerous in a control algorithm. The Cortex-M CPU supports saturated math instructions that stick at their maximum and minimum values.

While this is a problem for most applications, it is especially serious for applications such as motor control and safety critical applications. The Cortex-M3\M4 saturated math instructions prevent this kind of "roll round." When you use the saturated math instructions, if the variable reaches its maximum or minimum value it will stick (saturate) at that value. Once the variable is saturated the Q bit will be set. The Q bit is a "sticky" bit and must be cleared by the application code. The standard math instructions are not used by the C compiler by default. If you want to make use of the saturated math instructions, you have to access them by using compiler intrinsic or CMSIS-core functions.

```
uint32_t __SSAT(uint32_t value, uint32_t sat)
uint32_t __USAT(uint32_t value, uint32_t sat)
```

Interrupts and Multicycle Instructions

The next field in the PSR is the "ICI" and "IT" instruction flags. Most of the Cortex-M processor instructions are executed in a single cycle. However, some instructions such as load store multiple, multiply, and divide take multiple cycles. If an interrupt occurs while these instructions are executing, they have to be suspended while the interrupt is served. Once the interrupt has been served, we have to resume the multicycle instructions. The ICI field is managed by the Cortex-M processor so you do not need to do anything special in your application code. It does mean that when an exception is raised reaching the start of your interrupt routine will always take the same amount of cycles regardless of the instruction currently being executed by the CPU.

Conditional Execution—IF THEN Blocks

As we have seen in Chapter 1, the Cortex-M processors have a three-stage pipeline. This allows the fetch, decode, and execute units to operate in parallel greatly improving the performance of the processor. However, there is a disadvantage that every time the processor makes a jump, the pipeline has to be flushed and refilled. This introduces a big hit on performance as the pipeline has to be refilled before the execution of instructions can resume. The Cortex-M3 and Cortex-M4 reduce the branch penalty by having an instruction fetch unit that can carry out speculative branch target fetches from the possible branch address so the execution of the branch targets can start earlier. However, for small conditional branches, the Cortex-M processor has another trick up its sleeve. For a small conditional branch, for example

```
If(Y = = 0x12C){
  I++;
  }else{
  I-;}
```

which compiles to less than four instructions, the Cortex-M processor can compile the code as an IF THEN condition block. The instructions inside the IF THEN block are extended with a condition code. This condition code is compared to the state of the condition code flags in the PSR. If the condition matches the state of the flags, then the instruction will be executed and if it does not then the instruction will still enter the pipeline but will be executed as a no operation (NOP). This technique eliminates the branch and hence avoids the need to flush and refill the pipeline. So even though we are inserting NOP instructions, we still get better performance levels.

Table 3.1: Instruction Condition Codes

Condition Code	xPSR Flags Tested	Meaning
EQ	Z = 1	Equal
NE	Z = 0	Not equal
CS or HS	C = 1	Higher or same (unsigned)
CC or LO	C = 0	Lower (unsigned)
MI	N = 1	Negative
PL	N = 0	Positive or zero
VS	V = 1	Overflow
VC	V = 0	No overflow
HI	C = 1 and Z = 0	Higher (unsigned)
LS	C = 0 or Z = 1	Lower or same (unsigned)
GE	N = V	Greater than or equal (signed)
LT	N! = V	Less than (signed)
GT	Z = 0 and N = V	Greater than (signed)
LE	Z = 1 and N! = V	Less than or equal (signed)
AL	None	Always execute

To trigger an IF THEN block, we use the data processing instructions to update the PSR condition codes. By default, most instructions do not update the condition codes unless they have an S suffix added to the assembler opcode. This gives the compiler a great deal of flexibility in applying the IF THEN condition.

```
ADDS R1, R2, R3   //perform and add and set the xPSR flags
ADD R2, R4, R5   //Do some other instructions but do not modify the xPSR
ADD R5, R6, R7
IT VS   //IF THEN block conditional on the first ADD instruction
SUBVS R3, R2, R4
```

Consistently our 'C' IF THEN ELSE statement can be compiled to four instructions.

```
CMP     r6, #0x12C
ITE     EQ
STREQ   r4, [r0, #0x08]
STRNE   r5, [r0, #0x04]
```

The CMP compare instruction is used to perform the test and will set or clear the zero Z flag in the PSR. The IF THEN block is created by the IF THEN (IT) instruction. The IT instruction is always followed by one conditionally executable instruction and optionally up to four conditionally executable instructions. The format of the IT instruction is as follows:

```
IT x y z cond
```

The x, y, and z parameters enable the second, third, and fourth instructions to be a part of the conditional block. There can be a further THEN or ELSE instruction. The cond parameter is the condition applied to the first instruction. So,

```
ITTTE NE   A four-instruction IF THEN block with three THEN instructions which executes
           when Z = 1 followed by an ELSE instruction which executes when Z = 1
ITE GE     A two-instruction IF THEN block with one THEN instruction which executes when
           N = V and one ELSE instruction which executes when N! = V
```

The use of conditional executable IF THEN blocks is left up to the compiler. Generally, at low levels of optimization IF THEN blocks are not used; this gives a good debug view. However, at high levels of optimization, the compiler will make use of IF THEN blocks. So, normally there will not be any strange side effects introduced by the conditional execution technique but there are a few rules to bear in mind.

First, conditional code blocks cannot be nested; generally the compiler will take care of this rule. Secondly, you cannot use a GOTO statement to jump into a conditional code block. If you do make this mistake the compiler will warn you and not generate such illegal code.

Thirdly, the only time you will really notice execution of an IF THEN condition block is during debugging. If you single step the debugger through the conditional statement, the conditional code will appear to execute even if the condition is false. If you are not aware of the Cortex-M condition code blocks, this can be a great cause of confusion!

The next bit in the PSR is the T or Thumb bit. This is a legacy bit from the earlier ARM CPUs and is set to one when the processor is released from reset. If your code tries to clear this bit, it will be set to zero but a fault exception will also be raised. In previous CPUs, the T bit was used to indicate that the Thumb 16-bit instruction set was running. It is included in the Cortex-M PSR to maintain compatibility with earlier ARM CPUs and allow legacy 16-bit Thumb code to be executed on the Cortex-M processors. The final field in the PSR is the exception number field. The Cortex NVIC can support up to 256 exception sources. When an exception is being processed, the exception number is stored here. As we will see later, this field is not used by the application code when handling an interrupt, though it can be a useful reference when debugging.

Exercise: Saturated Math and Conditional Execution

In this exercise, we will use a simple program to examine the CPU registers and make use of the saturated math instructions. We will also rebuild the project to use conditional execution.

Open the project in C:\exercises\saturation.

```
int a,range = 300;   char c;
int main (void){
while (1){
for(a = 0;a < range;a++){
   c = a;
}}
```

This program increments an integer variable from 0 to 300 and copies it to a char variable.

Build the program and start the debugger.

Add the two variables to the logic analyzer and run the program.

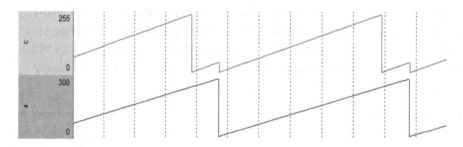

In the logic analyzer window, we can see that while the integer variable performs as expected the char variable saturates when it reaches 255 and "rolls over" to zero and begins incrementing again.

Stop the debugger and modify the code as shown below to use saturated math.

```
#define Q_FLAG 0x08000000
int a,range = 300;
char c;
unsigned int xPSR;

int main (void){
while (1){
for(a = 0;a < range;a++){
  c = __SSAT (a, 9);
  }
}}
```

This code replaces the equate statement with a saturated intrinsic function that saturates on the ninth bit of the integer value. This allows values 0−255 to be written to the byte value; any other values will saturate at the maximum allowable value of 255.

Build the project and start the debugger.

Run the code and view the contents of the variables in the logic analyzer.

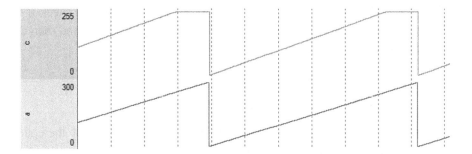

Now the char variable saturates rather than "rolling over." This is still wrong but not as potentially catastrophically wrong as the rollover case.

In the registers window, click on the xPSR regiset to view the flags.

In addition to the normal NVCZ condition code flags, the saturation Q bit is set.

Stop the debugger and modify the code as shown below.

```
#define Q_FLAG 0x08000000
int a, range = 300;   char c;   unsigned int APSR;
register unsigned int apsr __asm("apsr");
int main (void){
while (1){
for(a = 0;a < range;a++){
    c = __SSAT(a, 9);
    }
APSR = __get_APSR ();
if(APSR&Q_FLAG){
    range-;
```

```
}
apsr = apsr&~Q_FLAG;
}}
```

Once we have written to the char variable, it is possible to read the xPSR and check if the Q bit is set. If the variable has saturated, we can take some corrective action and then clear the Q bit for the next iteration.

Build the project and start the debugger.

Run the code and observe the variables in the watch window.

Now, when the data is over range, the char variable will saturate and gradually the code will adjust the range variable until the output data fits into the char variable.

Set breakpoints on lines 19 and 22 to enclose the Q bit test.

```
●19 ⊟if(xPSR&Q_FLAG){
  20 │     range--;
  21 ├}
●22 │  apsr = apsr&~Q_FLAG;
```

Reset the program and then run the code until the first breakpoint is reached.

Open the disassembly window and examine the code generated for the Q bit test.

```
    19: if(xPSR&Q_FLAG){
MOV   r0,r1
LDR   r0,[r0,#0x00]
TST   r0,#0x8000000
BEQ   0x080003E0
  20:     range-;
  21: }else{
LDR   r0,[pc,#44]; @0x08000400
LDR   r0,[r0,#0x00]
SUB   r0,r0,#0x01
LDR   r1,[pc,#36]; @0x08000400
STR   r0,[r1,#0x00]
B    0x080003E8
  22: locked=1;
  23: }
MOV   r0,#0x01
LDR   r1,[pc,#32]; @0x08000408
STR   r0,[r1,#0x00]
```

Also make a note of the value in the state counter. This is the number of cycles used since reset to reach this point.

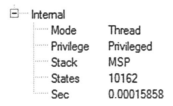

Now run the code until it hits the next breakpoint and again make a note of the state counter.

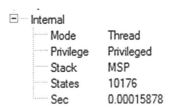

Stop the debugger and open the Options for Target\C tab.

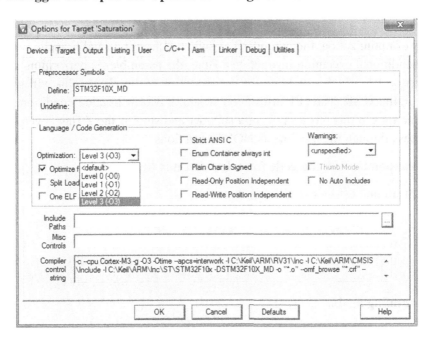

Change the optimization level from Level 0 to Level 3.

Close the Options for Target window and rebuild the project.

Now repeat the cycle count measurement by running to the two breakpoints.

⊟ Internal			⊟ Internal	
Mode	Thread		Mode	Thread
Privil...	Privileged		Privil...	Privileged
Stack	MSP		Stack	MSP
States	3480		States	3473
Sec	0.00006100		Sec	0.00006090

Now the Q bit test takes 7 cycles as opposed to the original 14.

Examine the disassembly code for the Q bit test.

```
 19: if(xPSR&Q_FLAG){
0x08000336 F0116F00 TST   r1,#0x8000000
 20:     range-;
 21: }else{
ITTE  NE
STRNE   r1,[r0,#0x08]
 22: locked=1;
 23: }
STREQ   r4,[r0,#0x04]
```

At higher levels of optimization, the compiler has switched from test and branch instructions to conditional execution instructions. Here, the assembler is performing a bitwise AND test on R1, which is holding the current value of the xPSR. This will set or clear the Z flag in the xPSR. The ITT instruction sets up a two-instruction conditional block. The instructions in this block perform a subtract and store if the Z flag is zero; otherwise they pass through the pipeline as NOP instructions.

Remove the breakpoints. Run the code for a few seconds then halt it.

Set a breakpoint on one of the conditional instructions.

```
0x0800033A BF1A      ITTE      NE
0x0800033C 1E69      SUBNE     r1,r5,#1
0x0800033E 6081      STRNE     r1,[r0,#0x08]
```

Start the code running again.

The code will hit the breakpoint even though the breakpoint is within an IF THEN statement that should no longer be executed. This is simply because the conditional instructions are always executed.

Cortex-M Memory Map and Busses

While each manufacturer's Cortex-M device has different peripherals and memory sizes, ARM has defined a basic memory template that all devices must adhere to. This provides a standard layout so all the vendor-provided memory and peripherals are located in the same blocks of memory.

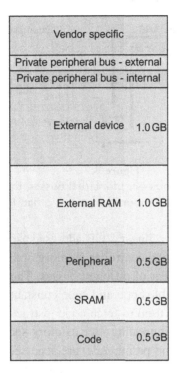

Figure 3.6
The Cortex-M memory map has a standard template which splits the 4 GB address range into specific memory regions. This memory template is common to all Cortex-M devices.

The Cortex-M memory template defines eight regions that cover the 4 GB address space of the Cortex-M processor. The first three regions are each 0.5 GB in size and are dedicated to the executable code space, internal SRAM, and internal peripherals. The next two regions are dedicated to external memory and memory mapped devices; both regions are 1 GB in size. The final three regions make up the Cortex-M processor memory space and contain the configuration registers for the Cortex-M processor and any vendor-specific registers.

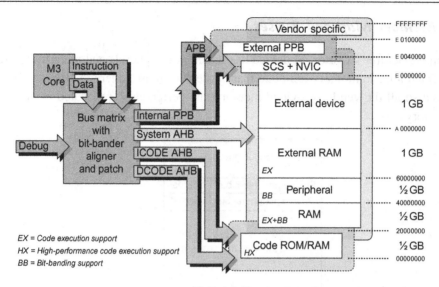

Figure 3.7

While the Cortex-M processor has a number of internal busses, these are essentially invisible to the software developer. The memory appears as a flat 4 GB address space.

While the Cortex-M memory map is a linear 4 GB address space with no paged regions or complex addressing modes, the microcontroller memory and peripherals are connected to the Cortex-M processor by a number of different busses. The first 0.5 GB of the address space is reserved for executable code and code constants. This region has two dedicated busses. The ICODE bus is used to fetch code instructions and the DCODE bus is used to fetch code constants. The remaining user memory spaces (internal SRAM and peripherals plus the external RAM and peripherals) are accessed by a separate system bus. The Cortex-M processor memory space has an additional private peripheral bus. While this may look complicated, as far as your application code is concerned you have one seamless memory space; the Cortex-M processor has separate internal busses to optimize its access to different memory regions.

As mentioned earlier, most of the instructions in the Thumb-2 instruction set are executed in a single cycle and the Cortex-M3 can run up to 200 MHz, and in fact some system on chip (SoC) designs manage to get the processor running even faster. However, current flash memory used to store the program has an access time of around 50 MHz. So there is a basic problem of pulling instructions out of the flash memory fast enough to feed the Cortex-M processor. When you are selecting a Cortex-M microcontroller it is important to study the data sheet to see how the silicon vendor has solved this problem. Typically, the flash memory will be arranged as 64- or 128-bit wide memory, so one read from the flash memory can load multiple instructions. These instructions are then held in a "memory accelerator" unit which then feeds the instructions to the Cortex-M processor as required.

The memory accelerator is a form of simple cache unit that is designed by the silicon vendor. Normally, this unit is disabled after reset, so you will need to enable it or the Cortex-M processor will be running directly from the flash memory. The overall performance of the Cortex-M processor will depend on how successfully this unit has been implemented by the designer of the microcontroller.

Write Buffer

The Cortex-M3 and Cortex-M4 contain a single entry data write buffer. This allows the CPU to make an entry into the write buffer and continue on to the next instruction while the write buffer completes the write to the real SRAM. If the write buffer is full, the CPU is forced to wait until it has finished its current write. While this is normally a transparent process to the application code, there are some cases where it is necessary to wait until the write has finished before continuing program execution. For example, if we are enabling an external bus on the microcontroller, it is necessary to wait until the write buffer has finished writing to the peripheral register and the bus is enabled before trying to access memory located on the external bus. The Cortex-M processor provides some memory barrier instructions to deal with these situations.

Memory Barrier Instructions

The memory barrier instructions halt execution of the application code until a memory write of an instruction has finished executing. They are used to ensure that a critical section of code has been completed before continuing execution of the application code.

Table 3.2: Memory Barrier Instructions

Instruction	Description
Data memory synchronization barrier (DMD)	Ensures all memory accesses are finished before a fresh memory access is made
Data synchronization barrier (DSB)	Ensures all memory accesses are finished before the next instruction is executed
Instruction synchronization barrier (ISB)	Ensures that all previous instructions are completed before the next instruction is executed. This also flushes the CPU pipeline

System Control Block

In addition to the CPU registers, the Cortex-M processors have a group of memory mapped configuration and status registers located near the top of the memory map starting at 0xE000 E008.

We will look at the key features supported by these registers in the rest of this book, but a summary is given below.

Table 3.3: The Cortex Processor Has Memory Mapped Configuration and Status Registers Located in the System Control Block (SCB)

Register	Size in Words	Description
Auxiliary control	1	Allows you to customize how some processor features are executed
CPU ID	1	Hardwired ID and revision numbers from ARM and the silicon manufacturer
Interrupt control and state	1	Provides pend bits for the systick and NMI Non Maskable interrupt along with extended interrupt pending\active information
Vector table offset	1	Programmable address offset to move the vector table to a new location in flash or SRAM memory
Application interrupt and reset control	1	Allows you to configure the PRIGROUP and generate CPU and microcontroller resets
System control	1	Controls configuration of the processor sleep modes
Configuration and control	1	Configures CPU operating mode and some fault exceptions
System handler priority	3	These registers hold the 8-bit priority fields for the configurable processor exceptions
System handler control and state	1	Shows the cause of a bus, memory management, or usage fault
Configurable fault status	1	Shows the cause of a bus, memory management, or usage fault
Hard fault status	1	Shows what event caused a hard fault
Memory manager fault address	1	Holds the address of the memory location that generated the memory fault
Bus fault address	1	Holds the address of the memory location that generated the memory fault

The Cortex-M instruction set has addressing instructions that allow you to load and store 8-, 16-, and 32-bit quantities. Unlike the ARM7 and ARM9, the 16- and 32-bit quantities do not need to be aligned on word or halfword boundaries. This gives the compiler and linker the maximum flexibility to fully pack the SRAM memory. However, there is a penalty to be paid for this flexibility because unaligned transfers take longer to carry out. The Cortex-M instruction set contains load and store multiple instructions that can transfer multiple registers to and from memory in one instruction. This takes multiple processor cycles but uses only one 2-byte or 4-byte instruction. This allows for very efficient stack manipulation and block memory copy. The load and store multiple instructions only work for word aligned data. So, if you use unaligned data, the compiler is forced to use multiple individual load and store instructions to achieve the same thing. While you are making full use of the valuable internal SRAM, you are potentially increasing the application instruction size. Unaligned data is for user data only; you must ensure that the stacks are word aligned. The main stack pointer's (MSP) initial value is determined by the linker, but the second stack pointer, the process stack pointer (PSP), is enabled and initialized by the user. We will look at using the PSP in Chapter 5.

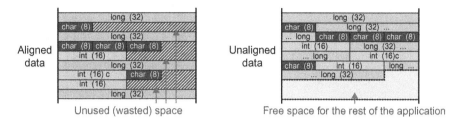

Figure 3.8
Unlike the earlier ARM7 and ARM9 CPUs, the Cortex processor can make unaligned memory accesses. This allows the compiler and linker to make the best use of the SRAM device.

Bit Manipulation

In a small embedded system, it is often necessary to set and clear individual bits within the SRAM and peripheral registers. By using the standard addressing instructions, we can set and clear individual bits by using the C language bitwise AND and OR commands. While this works fine, the Cortex-M processors provide a more efficient bit manipulation method.

The Cortex-M processor provides a method called "bit banding" which allows individual SRAM and peripheral register bits to be set and cleared in a very efficient manner.

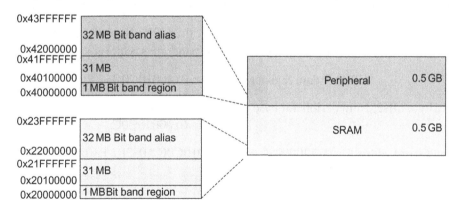

Figure 3.9
Bit banding is a technique that allows the first 1 MB of the SRAM region and the first 1 MB of the peripheral region to be bit addressed.

The first 1 MB of the SRAM region and the first 1 MB of the peripheral region are defined as bit band regions. This means that every memory location in these regions is bit addressable. So, in practice for today's microcontrollers, all of their internal SRAM and peripheral registers are bit addressable. Bit banding works by creating an alias word address for each bit of real memory or peripheral register bit. So, the 1 MB of real SRAM is aliased to 32 MBytes of virtual word addresses.

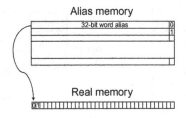

Figure 3.10
Each bit in the real memory is mapped to a word address in the alias memory.

This means that each bit of real RAM or peripheral register is mapped to a word address in the bit band alias region and by writing 1s and 0s to the alias word address we can set and clear the real memory bit location. Similarly, if we write a word to the real memory location, we can read the bit band alias address to check the current state of a bit in that word. To use bit banding, you simply need to calculate the word address in the bit band region that maps to the bit location in the real memory that you want to modify. Then create a pointer to the word address in the bit band region. Once this is done you can control the real bit memory location by reading and writing to the alias region via the pointer. The calculation for the word address in the bit band alias region is as follows:

Bit band word address = bit band base + (byte offset × 32) + (bit number × 4)

So, for example, if we want to read and write to bit 8 of the GPIO B port register on a typical Cortex microcontroller, we can calculate the bit band alias address as follows:

GPIO B data register address = 0x40010C0C

Register byte offset from peripheral base address = 0x40010C0C − 0x40000000
= 0x00010C0C

Bit band word address = 0x42000000 + (0x000010C0C * 0x20) + (0x8 * 0x4)
= 0x00010C0C

Now we can define a pointer to this address.

```
#define GPIO_PORTB_BIT8 = (*((volatile unsigned long *)0x422181A0)
```

Now by reading and writing to this word address, we can directly control the individual port bit.

```
GPIO_PORTB_BIT8 = 1   //set the port pin
```

This will compile the following assembler instructions:

```
Opcode     Assembler
F04F0001   MOV r0,#0x01
4927 LDR   r1,[pc,#156]; @0x080002A8
6008 STR   r0,[r1,#0x00]
```

This sequence uses one 32-bit instruction and two 16-bit instructions or a total of 8 bytes. If we compare this to setting the port pin by using a logical OR to write directly to the port register.

```
GPIOB->ODR |= 0x00000100;   //LED on
```

We then get the following code sequence:

```
Opcode     Assembler
481E LDR   r0,[pc,#120]; @0x080002AC
6800 LDR   r0,[r0,#0x00]
F4407080 ORR r0,r0,#0x100
491C LDR   r1,[pc,#112]; @0x080002AC
6008 STR   r0,[r1,#0x00]
```

This uses four 16-bit instructions and one 32-bit instruction or 12 bytes.

The use of bit banding gives us a win-win situation, smaller code size, and faster operation. So, as a simple rule, if you are going to repetitively access a single bit location you should use bit banding to generate the most efficient code.

You may find some compiler tools or silicon vendor software libraries that provide macro functions to support bit banding. You should generally avoid using such macros as they do not yield the most efficient code.

Exercise: Bit Banding

In this exercise, we will look at defining a bit band variable to toggle a port pin and compare its use to bitwise AND and OR instructions.

Open the project in c:\exercises\bitband.

In this exercise, we want to toggle an individual port pin. We will use the port B bit 8 pin as we have already done the calculation for the alias word address.

So, now in the C code, we can define a pointer to the bit band address.

```
#define PortB_Bit8 (*((volatile unsigned long *)0x422181A0))
```

And in the application code, we can set and clear the port pin by writing to this pointer.

```
PortB_Bit8 = 1;
PortB_Bit8 = 0;
```

Build the project and start the debugger.

Enable the timing analysis with the Debug\Execution Profiling\Show Time menu.

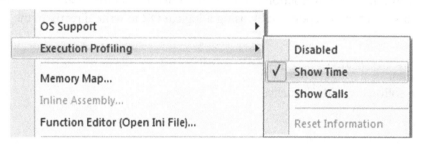

This opens an additional column in the debugger that will display execution time for each line. Compare the execution time for the bit band instruction compared to the AND and OR instructions.

```
   21              ⊟{
   22      0.069 us │ PortB_Bit8 = 1; //led on
   23
```

```
   30    0.153 us │ GPIOB->ODR |= 0x00000100;    //LED on
```

Open the disassemble window and examine the code generated for each method of setting the port pin.

The bit banding instructions are the best way to set and clear individual bits. In some microcontrollers, the silicon vendor has added support for fast bit manipulation on certain registers, typically GPIO ports, which is as fast as bit banding. Whatever methods are available, you should use them in any part of your code that repetitively manipulates a bit.

Dedicated Bit Manipulation Instructions

In addition to bit band support, the Thumb-2 instruction set has some dedicated bit orientated instructions. Some of these instructions are not directly accessible from the C language and are supported by compiler "intrinsic" calls. However, this is a moving target as over time the compiler tools are becoming increasingly optimized to make full use of the Thumb-2 instruction set. For example, the sign extend a byte (SXTB), sign extend a halfword (SXTH), zero extend a byte (UXTB), and zero extend a halfword (UXTH) instructions are used in data type conversions.

**Table 3.4: In Addition to Bit Banding the Cortex-M3 Processor
Has Some Dedicated Bit Manipulation Instructions**

BFC	Bit field clear
BFI	Bit field insert
SBFX	Signed bit field extract
SXTB	Sign extend a byte
SXTH	Sign extend a halfword
UBFX	Unsigned bit field extract
UXTB	Zero extend a byte
UXTH	Zero extend a halfword

Systick Timer

All of the Cortex-M processors also contain a standard timer. This is called the systick timer and is a 24-bit countdown timer with auto reload. Once started the systick timer will count down from its initial value. Once it reaches zero it will raise an interrupt and a new count value will be loaded from the reload register. The main purpose of this timer is to generate a periodic interrupt for an RTOS or other event-driven software. If you are not running an OS, you can also use it as a simple timer peripheral.

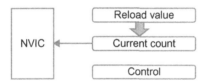

Figure 3.11
The systick timer is a 24-bit countdown timer with an auto reload. It is generally used to provide a periodic interrupt for an RTOS scheduler.

The default clock source for the systick timer is the Cortex-M CPU clock. It may be possible to switch to another clock source, but this will vary depending on the actual microcontroller you are using. While the systick timer is common to all the Cortex-M processors, its registers occupy the same memory locations within the Cortex-M3 and Cortex-M4. In the Cortex-M0 and Cortex-M0+, the systick registers are located in the SCB and have different symbolic names to avoid confusion. The systick timer interrupt line and all of the microcontroller peripheral lines are connected to the NVIC.

Nested Vector Interrupt Controller

Aside from the Cortex-M CPU, the next major unit within the Cortex-M processor is the NVIC. The usage of the NVIC is the same between all Cortex-M processors; once you have set up an interrupt on a Cortex-M3, the process is the same for the Cortex-M0, Cortex-M0+ , and Cortex-M4. The NVIC is designed for fast and efficient interrupt handling; on a Cortex-M3 you will reach the first line of C code in your interrupt routine after 12 cycles for a zero wait state memory system. This interrupt latency is fully deterministic so from any point in the background (noninterrupt) code you will enter the interrupt with the same latency. As we have seen, multicycle instructions can be halted with no overhead and then resumed once the interrupt has finished. On the Cortex-M3 and Cortex-M4, the NVIC supports up to 240 interrupt sources, and the Cortex-M0 can support up to 32 interrupt sources. The NVIC supports up to 256 interrupt priority levels on Cortex-M3 and Cortex-M4, and 4 priority levels on Cortex-M0.

Table 3.5: The NVIC Consists of Seven Register Groups That Allow You to Enable, Set Priority Levels for, and Monitor the User Interrupt Peripheral Channels

Register	Maximum Size in Words*	Description
Set enable	8	Provides an interrupt enable bit for each interrupt source
Clear enable	8	Provides an interrupt disable bit for each interrupt source
Set pending	8	Provides a set pending bit for each interrupt source
Clear pending	8	Provides a clear pending bit for each interrupt source
Active	8	Provides an interrupt active bit for each interrupt source
Priority	60	Provides an 8-bit priority field for each interrupt source
Software trigger	1	Write the interrupt channel number to generate a software interrupt

*The actual number of words used will depend on the number of interrupt channels implemented by the microcontroller manufacturer.

Operating Modes

While the Cortex CPU is executing background (noninterrupt code) code, the CPU is in an operating mode called thread mode. When an interrupt is raised, the NVIC will cause the processor to jump to the appropriate interrupt service routine (ISR). When this happens, the CPU changes to a new operating mode called handler mode. In simple applications without an OS, you can use the default configuration of the Cortex-M processor out of reset; there is no major functional difference in these operating modes and they can be ignored. The Cortex-M processors can be configured with a more complex operating model that introduces operating differences between thread and handler modes and we will look at this in Chapter 5.

Interrupt Handling—Entry

When a microcontroller peripheral raises an interrupt line, the NVIC will cause two things to happen in parallel. First, the exception vector is fetched over the ICODE bus. This is the address of the entry point into the ISR. This address is pushed into R15, the program counter, forcing the CPU to jump to the start of the interrupt routine. In parallel with this, the CPU will automatically push key registers onto the stack. This stack frame consists of the following registers; xPSR, PC, LR, R12, R3, R2, R1, R0. This stack frame preserves the state of the processor and provides R0–R3 for use by the ISR. If the ISR needs to use more CPU registers, it must PUSH them onto the stack and POP them on exit.

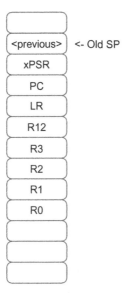

Figure 3.12
When an interrupt or exception occurs, the CPU will automatically push a stack frame. This consists of the xPSR, PC, LR, R12, and registers R0–R3. At the end of the interrupt or exception, the stack frame is automatically unstacked.

The interrupt entry process takes 12 cycles on the Cortex-M3\M4 and 16 cycles on the Cortex-M0. All of these actions are handled by microcode in the CPU and do not require any dedicated entry instructions like Long Jump (LJMP) or PUSH to be part of the application code.

The exception vectors are stored in an interrupt vector table. The interrupt vector table is located at the start of the address space; the first four bytes are used to hold the initial value of the stack pointer. The starting value of the stack pointer is calculated by the compiler and linker and automatically loaded into R13 when the Cortex processor is reset. The interrupt vector table then has address locations for every four bytes growing upward through the address space. The vector table holds the address of the ISR for each of the possible interrupt sources within the microcontroller. The vector table for each microcontroller comes predefined as part of the startup code. A label for each ISR is stored at each interrupt vector location. To create your ISR, you simply need to declare a void C function using the same name as the interrupt vector label.

```
      AREA  RESET, DATA, READONLY
      EXPORT __Vectors
 __Vectors  DCD  __initial_sp    ; Top of Stack
      DCD  Reset_Handler    ; Reset Handler
  ; External Interrupts
      DCD  WWDG_IRQHandler    ; Window Watchdog
      DCD  PVD_IRQHandler    ; PVD through EXTI Line detect
      DCD  TAMPER_IRQHandler    ; Tamper
      DCD  RTC_IRQHandler    ; RTC
      DCD  FLASH_IRQHandler    ; Flash
      DCD  RCC_IRQHandler    ; RCC
```

So, to create the C routine to handle an interrupt from the real-time clock, we create a C function named as follows:

```
void RTC_IRQHandler(void) {
 .......
 }
```

When the project is built, the linker will resolve the address of the C routine and locate it in the vector table in place of the label. If you are not using this particular interrupt in your project, the label still has to be declared to prevent an error occurring during the linking process. Following the interrupt vector table, there is a second table that declares all of the ISR addresses. These are declared as WEAK labels. This means that this declaration can be overwritten if the label is declared elsewhere in the project. In this case, they act as a "backstop" to prevent any linker errors if the interrupt routine is not formally declared in the project source code.

```
      EXPORT  WWDG_IRQHandler [WEAK]
      EXPORT PVD_IRQHandler  [WEAK]
      EXPORT TAMPER_IRQHandler  [WEAK]  DCD  EXTIO_IRQHandler  ; EXTI Line 0
```

Interrupt Handling—Exit

Once the ISR has finished its task, it will force a return from the interrupt to the point in the background code from where it left off. However, the Thumb-2 instruction set does not have a return or return from interrupt instruction. The ISR will use the same return method as a noninterrupt routine, namely a branch on R14, the link register. During normal operation, the link register will contain the correct return address. However, when we entered the interrupt, the current contents of R14 were pushed onto the stack and in their place the CPU entered a special code. When the CPU tries to branch on this code instead of doing a normal branch, it is forced to restore the stack frame and resume normal processing.

Table 3.6: At the Start of an Exception or Interrupt R14 (Link Register) Is Pushed Onto the Stack

Interrupt Return Value	Meaning
0xFFFFFFF9	Return to thread mode and use the MSP
0xFFFFFFFD	Return to thread mode and use the PSP
0xFFFFFFF1	Return to handler mode

The CPU then places a control word in R14. At the end of the interrupt, the code will branch on R14. The control word is not a valid return address and will cause the CPU to retrieve a stack frame and return to the correct operating mode.

Interrupt Handling—Exit: Important!

The interrupt lines that connect the user peripheral interrupt sources to the NVIC interrupt channels can be level sensitive or edge sensitive. In many microcontrollers, the default is level sensitive. This means that once an interrupt has been raised, it will be asserted on the NVIC until it is cleared. This means that if you exit an ISR with the interrupt still asserted on the NVIC, a new interrupt will be raised. To cancel the interrupt, you must clear the interrupt status flags in the user peripheral before exiting the ISR. If the peripheral generates another interrupt while its interrupt line is asserted, a further interrupt will not be raised. If you clear the interrupt status flags at the beginning of the interrupt routine, then any further interrupts from the peripheral will be served. To further complicate things, some peripherals will automatically clear some of their status flags. For example, an ADC conversion complete flag may be automatically cleared when the ADC results register is read. Keep this in mind when you are reading the microcontroller user manual.

Exercise: Systick Interrupt

This project demonstrates setting up an interrupt using the systick timer.

Open the project in c:\exercises\exception.

This application consists of the minimum amount of code necessary to get the Cortex-M processor running and to generate a systick interrupt.

Open the main.c file.

```
#include "stm32f10x.h"
#define SYSTICK_COUNT_ENABLE  1
#define SYSTICK_INTERRUPT_ENABLE  2
int main (void)
{
GPIOB->CRH = 0x33333333;
SysTick->VAL = 0x9000;
SysTick->LOAD = 0x9000;
SysTick->CTRL = SYSTICK_INTERRUPT_ENABLE | SYSTICK_COUNT_ENABLE;
while(1);
}
```

The main function configures a bank of port pins as outputs. Next, we load the systick timer and reload the register and then enable the timer and its interrupt line to the NVIC. Once this is done the background code sits in a while loop doing nothing.

When the timer counts down to zero, it will generate an interrupt that will run the systick ISR.

```
void SysTick_Handler (void)
{
static unsigned char count = 0;
if(count++ > 0x60){
GPIOB->ODR ^= 0xFFFFFFFF;
count = 0;
}}
```

The interrupt routine is then used to periodically toggle the GPIO lines.

Open the STM32F10x.s file and locate the vector table.

```
SysTick_Handler PROC
  EXPORT SysTick_Handler  [WEAK]
  B  .
  ENDP
```

The vector table provides standard labels for each interrupt source created as "weak" declarations. To create a C ISR, we simply need to use the label name as the name for a void function. The C function will then override the assembled stub and be called when the interrupt is raised.

Build the project and start the debugger.

Without running the code, open the register window and examine the state of the registers.

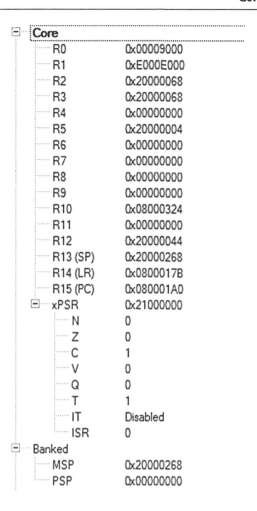

In particular, note the value of the stack pointer (R13), the link register (R14), and the PSR.

Set a breakpoint in the interrupt routine and start running the code.

```
26  void SysTick_Handler ( void)
27 ⊟ {
28    static unsigned char count = 0;
29      if (count++>0x60)
30 ⊟ {
31    GPIOB->ODR ^=0xFFFFFFFF;
32    count = 0;
33    }
34    }
```

When the code hits the breakpoint again examine the register window.

               R13 (SP)   0x20000248
               R14 (LR)   0xFFFFFFF9

Now, R14 has the interrupt return code in place of a normal return address and the stack pointer has been decremented by 32 bytes.

Open a memory window at 0x20000248 and decode the stack frame.

Address: 0x20000248

0x20000248: 00000003 E000E000 20000068 20000068 20000044 0800017B 080001A6 21000000 00000000

Now open the Peripherals\Core Peripherals\Nested Vectored Interrupt Controller menu.

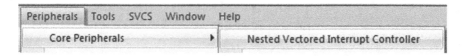

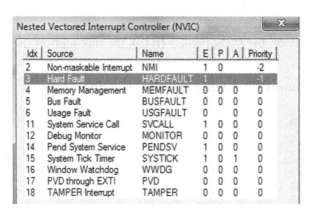

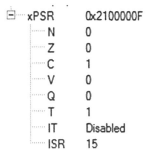

The NVIC peripheral window shows the state of each interrupt line. Line 15 is the systick timer and it is enabled and active (P = pending). This also ties up with the ISR channel number in the PSR.

Now set a breakpoint on the closing brace of the interrupt function and run the code.

```
26   void SysTick_Handler ( void)
27 ⊟ {
28   │ static unsigned char count = 0;
29   │    if(count++>0x60)
30 ⊟ {
31   │ GPIOB->ODR ^=0xFFFFFFFF;
32   │ count = 0;
33   ├ }
34   │ }
```

Now open the disassembly window and view the return instruction.

```
      32: count = 0;
      33: }
x080001C6 2000        MOVS        r0,#0x00
x080001C8 4611        MOV         r1,r2
x080001CA 7008        STRB        r0,[r1,#0x00]
      34: }
x080001CC 4770        BX          lr
```

The return instruction is a branch instruction, same as if you were returning from a subroutine. However, the code in the link register (R14) will force the CPU to unstack and return from the interrupt. Single step this instruction (F11) and observe the return to the background code and restoration of the stacked values to the CPU registers.

Cortex-M Processor Exceptions

In addition to the peripheral interrupt lines, the Cortex-M processor has some internal exceptions and these occupy the first 15 locations of the vector table.

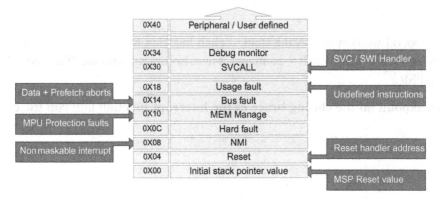

Figure 3.13
The first four bytes of memory hold the initial stack value. The vector table starts from 0x00000004. The first 10 vectors are for the Cortex processor while the remainder are for user peripherals.

The first location in the vector table is the reset handler. When the Cortex-M processor is reset, the address stored here will be loaded into the Cortex-M program counter forcing a jump to the start of your application code. The next location in the vector table is for a nonmaskable interrupt. How this is implemented will depend on the specific microcontroller you are using. It may, for example, be connected to an external pin on the microcontroller or to a peripheral such as a watchdog within the microcontroller. The next four exceptions are for handling faults that may occur during execution of the application code. All of these exceptions are present on the Cortex-M3 and Cortex-M4 but only the hard fault handler is implemented on the Cortex-M0. The types of faults that can be detected by the processor are usage fault, bus fault, memory manager fault, and hard fault.

Usage Fault

A usage fault occurs when the application code has incorrectly used the Cortex-M processor. The typical cause is when the processor has been given an invalid opcode to execute. Most ARM compilers can generate code for a range of ARM processor cores.

So, it is possible to incorrectly configure the compiler and to produce code that will not run on a Cortex-M processor. Other causes of a usage fault are shown below.

Table 3.7: Possible Causes of the Usage Fault Exception

Undefined instruction
Invalid interrupt return address
Unaligned memory access using load and store multiple instructions
Divide by zero*
Unaligned memory access*

*This feature must be enabled in the SCB configurable fault usage register.

Bus Fault

A bus fault is raised when an error is detected on the AHB bus matrix (see more about the bus matrix in Chapter 5). The potential reasons for this fault are as follows.

Table 3.8: Possible Causes of the Bus Fault Exception

Invalid memory region
Wrong size of data transfer, that is, a byte write to a word-only peripheral register
Wrong processor privilege level (we will look at privilege levels in Chapter 5)

Memory Manager Fault

The MPU is an optional Cortex-M processor peripheral that can be added when the microcontroller is designed. It is available on all variants except the Cortex-M0. The MPU is used to control access to different regions of the Cortex-M address space depending on the operating mode of the processor. This will be looked at in more detail in Chapter 5. The MPU will raise an exception in the following cases.

Table 3.9: Possible Causes of the Memory Manager Fault Exception

Accessing an MPU region with the wrong privilege level
Writing to a read-only region
Accessing a memory location outside of the defined MPU regions
Program execution from memory region that is defined as nonexecutable

Hard Fault

A hard fault can be raised in two ways. First, if a bus error occurs when the vector table is being read. Secondly, the hard fault exception is also reached through fault escalation. This means that if the usage, memory manager, or bus fault exceptions are disabled, or if the exception service does not have sufficient priority level, then the fault will escalate to a hard fault.

Enabling Fault Exceptions

The hard fault handler is always enabled and can only be disabled by setting the CPU FAULTMASK register. The other fault exceptions must be enabled in the SCB, system handler control (SHC), and state register (SR) (SCB->SHCSR). The SCB->SHCSR register also contains pend and active bits for each fault exception.

We will look at the fault exceptions and tracking faults in Chapter 8.

Priority and Preemption

The NVIC contains a group of priority registers with an 8-bit field for each interrupt source. In its default configuration, the top 7 bits of the priority register allow you to define the preemption level. The lower the preemption level, the more important the interrupt. So, if an interrupt is being served and a second interrupt is raised with a lower preemption level, then the state of the current interrupt will be saved and the processor will serve the new interrupt. When it is finished, the processor will resume serving the original interrupt provided a higher-priority interrupt is not pending.

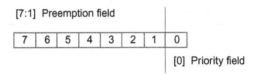

Figure 3.14
Each peripheral priority register consists of a configurable preemption field and a subpriority field.

The least significant bit is the subpriority bit. If two interrupts are raised with the same preemption level, the interrupt with the lowest subpriority level will be served first. This means we have 128 preemption levels each with two subpriority levels.

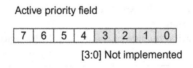

Figure 3.15
Each priority register is 8-bits wide. However, the silicon manufacturer may not implement all of the priority bits. The implemented bits always extend from the most significant bits (MSB) toward the least significant bits (LSB).

When the microcontroller is designed, the manufacturer can define the number of active bits in the priority register. For the Cortex-M3 and Cortex-M4, this can be a minimum of three and up to a maximum of eight. For the Cortex-M0, Cortex-M0+, and Cortex-M1, it is always 2 bits. Reducing the number of active priority bits reduces the NVIC gate count and hence its power consumption. If the manufacturer does not implement the full 8 bits of the priority register, the LSB will be disabled. This makes it safer to port code between microcontrollers with different numbers of active priority bits. You will need to check the manufacturer's datasheet to see how many bits of the priority register are active.

Groups and Subgroups

By default, in NVIC the first 7 bits of the priority register define the preemption level and the LSB defines the subpriority level. This split between preemption group and priority subgroup can be modified by writing to the NVIC priority group field in the application interrupt and reset control (AIRC) register. This register allows us to change the size of the preemption group field and priority subgroup. On reset, this register defaults to priority group zero.

Table 3.10: Priority Group and Subgroup Values

Priority Group	Preempt Group Bits	Subpriority Group Bits
0	7−1	0
1	7−2	1−0
2	7−3	2−0
3	7−4	3−0
4	7−5	4−0
5	7−6	5−0
6	7	6−0
7	None	7−0

So, for example, if our microcontroller has four active priority bits we could select priority group 5, which would give us four levels of preemption each with four levels of subpriority.

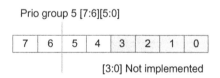

Prio group 5 [7:6][5:0]

| 7 | 6 | 5 | 4 | 3 | 2 | 1 | 0 |

[3:0] Not implemented

Figure 3.16
A priority register with four active bits and priority group 5. This yields four preempt levels and four priority levels.

The highest preemption level for a user exception is zero, however, some of the Cortex-M processor exceptions have negative priority levels so they will always preempt a user interrupt.

	Exception	Name	Priority	Descriptions
Fault Mode and Startup Handlers	1	Reset	−3 (Highest)	Reset
	2	NMI	−2	Nonmaskable Interrupt
	3	Hard fault	−1	Default fault if other hander not implemented
	4	Memory manage fault	Programmable	MPU violation or access to illegal locations
	5	Bus fault	Programmable	Fault if AHB interface receives error
	6	Usage fault	Programmable	Exceptions due to program errors
System Handlers	11	SVCall	Programmable	System service call
	12	Debug monitor	Programmable	Breakpoints, watch points, external debug
	14	PendSV	Programmable	Pendable service request for System Device
	15	Systick	Programmable	System Tick Timer
Custom Handlers	16	Interrupt #0	Programmable	External interrupt #0
	...	...	...	...
	...	...	...	...
	...	...	...	...
	255	Interrupt #239	Programmable	External interrupt #239

Figure 3.17

Cortex-M processor exceptions and possible priority levels.

Run Time Priority Control

There are three CPU registers that may be used to dynamically disable interrupt sources within the NVIC. These are the PRIMASK, FAULTMASK, and BASEPRI registers.

Table 3.11: The CPU PRIMASK, FAULTMASK, and BASEPRI Registers Are Used to Dynamically Disable Interrupts and Exceptions

CPU Mask Register	Description
PRIMASK	Disables all exceptions except hard fault and NMI
FAULTMASK	Disables all exceptions except NMI
BASEPRI	Disables all exceptions at the selected preemption level and lower preempt level

These registers are not memory mapped; they are CPU registers and may only be accessed with the MRS and MSR instructions. When programming in C, they may be accessed by dedicated compiler intrinsic instructions; we will look at these intrinsics more closely in Chapter 4.

Exception Model

When the NVIC serves a single interrupt, there is a delay of 12 cycles until we reach the ISR and a further 10 cycles at the end of the ISR until the Cortex-M processor resumes execution of the background code. This gives us fast deterministic handling of interrupts in a system that may have only one or two active interrupt sources.

Figure 3.18
When an exception is raised, a stack frame is pushed in parallel with the ISR address being fetched from the vector table. On the Cortex-M3 and Cortex-M4 this is always 12 cycles. On the Cortex-M0, it takes 16 cycles. The Cortex-M0+ takes 15 cycles.

In more complex systems, there may be many active interrupt sources all demanding to be served as efficiently as possible. The NVIC has been designed with a number of optimizations to ensure optimal interrupt handling in such a heavily loaded system. All of the interrupt handling optimizations described below are an integral part of the NVIC and as such are performed automatically by the NVIC and do not require any configuration by the application code.

NVIC Tail Chaining

In a very interrupt-driven design, we can often find that while the CPU is serving a high-priority interrupt a lower-priority interrupt is also pending. In the earlier ARM CPUs and many other processors, it was necessary to return from the interrupt by POPing the CPU context from the stack back into the CPU registers and then performing a fresh stack PUSH before running the pending ISR. This is quite wasteful in terms of CPU cycles as it performs two redundant stack operations.

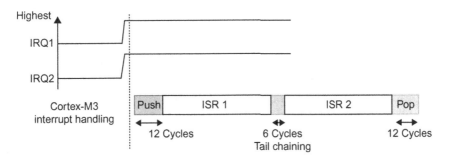

Figure 3.19
If an interrupt ISR is running and a lower-priority interrupt is raised, it will be automatically "tail chained" to run six cycles after the initial interrupt has terminated.

When this situation occurs on a Cortex-M processor, the NVIC uses a technique called tail chaining to eliminate the unnecessary stack operations. When the Cortex processor reaches the end of the active ISR and there is a pending interrupt, the NVIC simply forces the processor to vector to the pending ISR. This takes a fixed six cycles to fetch the start address of the pending interrupt routine and then execution of the next ISR can begin. Any further pending interrupts are dealt with in the same way. When there are no further interrupts pending, the stack frame will be POPed back to the processor registers and the CPU will resume execution of the background code. As you can see from Figure 3.19, tail chaining can significantly improve the latency between interrupt routines.

NVIC Late Arriving

Another situation that can occur is a "late arriving" high-priority interrupt. In this situation, a low-priority interrupt is raised followed almost immediately by a high-priority interrupt. Most microcontrollers will handle this by preempting the initial interrupt. This is undesirable because it will cause two stack frames to be pushed and delay the high-priority interrupt.

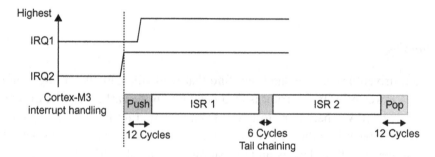

Figure 3.20
If the Cortex-M processor is entering the ISR and a higher-priority interrupt is raised, the NVIC will automatically switch to serve the high-priority interrupt. This will only happen if the initial interrupt is in its first 12 cycles.

If this situation occurs on a Cortex-M processor and the high-priority interrupt arrives within the initial 12-cycle PUSH of the low-priority stack frame, then the NVIC will switch to serving the high-priority interrupt and the low-priority interrupt will be tail chained to execute once the high-priority interrupt is finished. For the "late arriving" switch to happen, the high-priority interrupt must occur in the initial 12-cycle period of the low-priority interrupt. If it occurs any later than this, then it will preempt the low-priority interrupt, which requires the normal stack PUSH and POP.

NVIC POP Preemption

The final optimization technique used by the NVIC is called POP preemption. This is kind of a reversal of the late arriving technique discussed above.

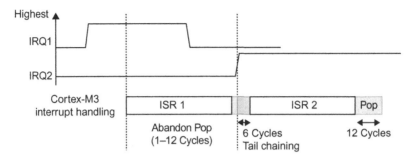

Figure 3.21
If an interrupt is raised while it is in its exiting 12 cycles, the processor will "rewind" the stack and serve the new interrupt with a minimum delay of six cycles.

When a typical microcontroller reaches the end of an ISR, it always has to restore the stack frame regardless of any pending interrupts. As we have seen above, the NVIC will use tail chaining to efficiently deal with any currently pending interrupts. However, if there are no pending interrupts, the stack frame will be restored to the CPU registers in the standard 12 cycles. If during this 12-cycle period a new interrupt is raised, the POPing of the stack frame will be halted and the stack pointer will be wound back to the beginning of the stack frame, the new interrupt vector will be fetched, and the new ISR will be executed. At the end of the new interrupt routine, we return to the background code through the usual 12-cycle POP process. This technique is called POP preemption.

It is important to remember that these three techniques, tail chaining, late arriving, and POP preemption, are all handled by the NVIC without any instructions being added to your application code.

Exercise: Working with Multiple Interrupts

This exercise extends our original systick exception exercise to enable a second ADC interrupt. We can use these two interrupts to examine the behavior of the NVIC when it has multiple interrupt sources.

Open the project in c:\exercises\multiple exceptions.

```
volatile unsigned char BACKGROUND = 0;unsigned char ADC = 0;unsigned char SYSTICK = 0;
int main (void){
int i;
```

```
GPIOB->CRH = 0x33333333;    //Configure the Port B LED pins
SysTick->VAL = 0x9000;    //Start value for the systick counter
SysTick->LOAD = 0x9000;    //Reload value
SysTick->CTRL = SYSTICK_INTERRUPT_ENABLE | SYSTICK_COUNT_ENABLE;
init_ADC();    //setup the ADC peripheral
ADC1->CR1 |= (1UL << 5);    // enable for EOC Interrupt
NVIC->ISER[0] = (1UL << 18);    // enable ADC Interrupt
ADC1->CR2 |= (1UL << 0);    // ADC enable
while(1){
  BACKGROUND = 1;
}}
```

We initialize the systick timer in the same manner as before. In addition, the ADC peripheral is also configured. To enable a peripheral interrupt, it is necessary to enable the interrupt source in the peripheral and also enable its interrupt channel in the NVIC by setting the correct bit in the NVIC Interrupt Set Enable Register ISER registers.

We have also added three variables: BACKGROUND, ADC, and SYSTICK. These will be set to logic one when the matching region of code is executing and zero at other times. This allows us to track execution of each region of code using the debugger logic analyzer.

```
void ADC_IRQHandler (void){
int i;
BACKGROUND    = 0;
SYSTICK    = 0;
for (i = 0;i <0x1000;i++){
  ADC = 1;
}
ADC1->SR &= ~(1 << 1);    /* clear EOC interrupt    */
ADC = 0;
}
```

The ADC interrupt handler sets the execution region variables then sits in a delay loop. Before exiting it also writes to the ADC status register to clear the end of conversion flag. This deasserts the ADC interrupt request to the NVIC.

```
void SysTick_Handler (void){
int i;
BACKGROUND = 0;
ADC = 0;
ADC1->CR2 |= (1UL << 22);
for (i = 0;i <0x1000;i++){
```

```
    SYSTICK=1;
  }
  SYSTICK =0;
  }
```

The systick interrupt handler is similar to the ADC handler. It sets the region execution variables and sits in a delay loop before exiting. It also writes to the ADC control register to trigger a single ADC conversion.

Build the project and start the simulator.

Add each of the execution variables to the logic analyzer and start running the code.

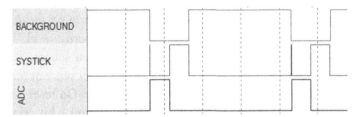

The systick interrupt is raised, which starts the ADC conversion. The ADC finishes conversion and raises its interrupt before the systick interrupt completes so it enters a PEND state. When the systick interrupt completes, the ADC interrupt is tail chained and begins execution without returning to the background code.

Exit the debugger and comment out the line of code that clears the ADC at the end of conversion flag.

```
//ADC1->SR &= ~(1 << 1);
```

Build the code and restart the debugger and observe the execution of the interrupts in the logic analyzer window.

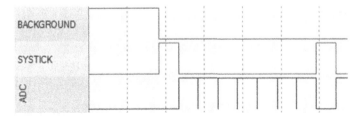

After the first ADC interrupt has been raised, the ADC 'End Of Conversion' interrupt status flag will be set and the ADC interrupt line to the NVIC stays asserted. If you quit the interrupt service routine without clearing all the pending interrupt status flags continuous

ADC interrupts to be raised by the NVIC, blocking the activity of the background code. The systick interrupt has the same priority as the ADC so it will be tail chained to run after the current ADC interrupt has finished. Neglecting to clear interrupt status flags is the most common mistake made when first starting to work with the Cortex-M processors.

Exit the debugger and uncomment the end of conversion code.

```
ADC1->SR &= ~(1 << 1);
```

Add the following lines to the background initializing code.

```
NVIC->IP[18] = (2 << 4);
SCB->SHP[11] = (3 << 4);
```

This programs the user peripheral NVIC interrupt priority registers to set the ADC priority level and the system handler priority registers to set the systick priority level. These are both byte arrays that cover the 8-bit priority field for each exception source. However, on this microcontroller the manufacturer has implemented four priority bits out of the possible eight. The priority bits are located in the upper nibble of each byte. On reset, the PRIGROUP is set to zero, which creates a 7-bit preemption field and 1-bit priority field.

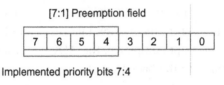

Figure 3.22
After reset, a microcontroller with four implemented priority bits will have 16 levels of preemption.

On our device all of the available priority bits are located in the preemption field giving us 16 levels of priority preemption.

Build the code, restart the debugger, and observe the execution of the interrupts in the logic analyzer window.

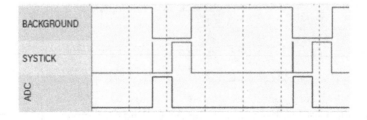

The ADC now has the highest preemption value so as soon as its interrupt is raised, it will preempt the systick interrupt. When it completes, the systick interrupt will resume and complete before returning to the background code.

Exit the debugger and uncomment the following lines in the background initialization code.

The AIRC register cannot be written too freely. It is protected by a key field that must be programmed with the value 0x5FA before a write is successful.

```
temp = SCB->AIRCR;
temp &= ~(SCB_AIRCR_VECTKEY_Msk | SCB_AIRCR_PRIGROUP_Msk);
temp = (temp|((uint32_t)0x5FA << 16) | (0x05 << 8));
SCB->AIRCR = temp;
```

This programs the PRIGROUP field in the AIRC register to a value of 5, which means a 2-bit preemption field and a 6-bit priority field. This maps onto the available 4-bit priority field giving four levels of preemption each with four levels of priority.

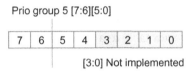

Prio group 5 [7:6][5:0]

| 7 | 6 | 5 | 4 | 3 | 2 | 1 | 0 |

[3:0] Not implemented

Figure 3.23
When the PRIGROUP field in the AIRC register is set to 5 each priority register has a 2-bit preemption field and a 6-bit priority field.

Build the code, restart the debugger, and observe the execution of the interrupts in the logic analyzer window.

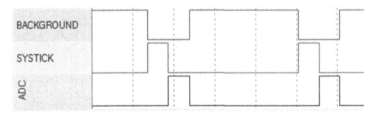

The ADC interrupt is no longer preempting the systick timer despite them having different values in their priority registers. This is because they now have different values in the priority field but with the same preempt value.

Exit the debugger and change the interrupt priorities as shown below.

```
NVIC->IP[18] = (2<<6 | 2<<4);
SCB->SHP[11] = (1<<6 | 3<<4);
```

Set the base priority register to block the ADC preempt group.

```
__set_BASEPRI ( 2<<6 );
```

Build the code, restart the debugger, and observe the execution of the interrupts in the logic analyzer window.

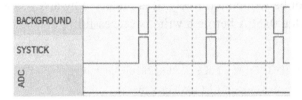

Setting the BASEPRI register has disabled the ADC interrupt and any other interrupts that are on the same level of preempt group or lower.

Bootloader Support

While the interrupt vector table is located at the start of memory when the Cortex-M processor is reset, it is possible to relocate the vector table to a different location in memory. As software embedded in small microcontrollers becomes more sophisticated, there is an increasing need to develop systems with a permanent bootloader program that can check the integrity of the main application code before it runs and checks for a program update that can be delivered by various serial interfaces (e.g., Ethernet, USB, UART) or an SD/multimedia card.

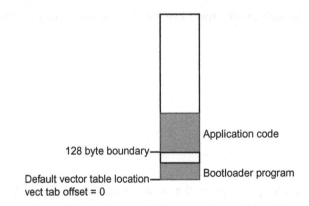

Figure 3.24
A bootloader program can be placed into the first sector in the flash memory. It will check if there is an update to the application code before starting the main application program.

Once the bootloader has performed its checks and, if necessary, updated the application program, it will jump the program counter to the start of the application code, which will start running. To operate correctly, the application code requires the vector table to be mapped to the start address of the application code.

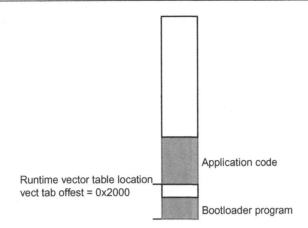

Figure 3.25
When the application code starts to run, it must relocate the vector table to the start of the application code by programming the NVIC vector table offset register.

The vector table can be relocated by writing to a register in the SCB called the vector table offset register. This register allows you to relocate the vector table to any 128-byte boundary in the Cortex processor memory map.

Exercise: Bootloader

This exercise demonstrates how a bootloader and application program can both be resident on the same Cortex-M microcontroller and how to debug such a system.

Open the multiworkspace project in C:\exercises\bootloader.

This is a more advanced feature of the μVision IDE that allows you to view two or more projects at the same time.

The workspace consists of two projects, the bootloader project, which is built to run on the Cortex processor reset vector as normal, and the blinky project, which is our application. First, we need to build the blinky project to run from an application address that is not in the same flash sector as the bootloader. In this example, the application address is chosen to be 0x2000.

Expand the blinky project, right click on the workspace folder, and set it as the active project.

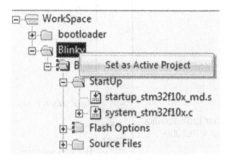

Now click on the blinky project folder and open the Options for Target\Target tab.

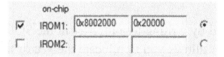

The normal start address for this chip is 0x8000000 and we have increased this to 0x8002000.

Open the system_stm32F10x.c file and locate line 128.

```
#define VECT_TAB_OFFSET 0x02000
```

This contains a define for the vector table offset register. Normally, this is zero but if we set this to 0x2000 the vector table will be remapped to match our application code when it starts running.

Build the blinky project.

Expand the bootloader project and set it as the active project.

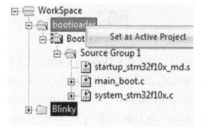

Open main_boot.c.

The bootloader program demonstrates how to jump from one program to the start of another. We need to define the start address of our second program. This must be a multiple of 128 bytes (0x200). In addition, a void function pointer is also defined.

```
#define APPLICATION_ADDRESS 0x2000
typedef void (*pFunction)(void);
pFunction Jump_To_Application;
uint32_t JumpAddress;
```

When the bootloader code enters main, it performs a custom check on the application code flash, such as a checksum, and could also test other critical aspects of the hardware. The bootloader then checks to see if there is a new application ready to be programmed into the application area. This could be in response to a command from an upgrade utility via a serial interface, for example. If the application program checks fail or a new update is available, we enter into the main bootloader code.

```
int main(void) {
uint32_t bootFlags;
    /* check the integrity of the application code */
    /* check if an update is available */
    /* if either case is true set a bit in the bootflags register */
    bootFlags = 0;
    if (bootFlags! = 0) {
    //enter the flash update code here
    }
```

If the application code and the hardware is OK, then the bootloader will handover to the application code. The reset vector of the application code is now located at the application address + 4. This can be loaded into the function pointer, which can then be executed resulting in a jump to the start of the application code. Before we jump to the application code, it is also necessary to load the stack pointer with the start address expected by the application code.

```
else {
    JumpAddress = *(__IO uint32_t*) (APPLICATION_ADDRESS + 4);
    Jump_To_Application = (pFunction) JumpAddress;
    // read the first four bytes of the application code and program this value into the
        stack pointer;
    //This sets the stack ready for the application code
    __set_MSP(*(__IO uint32_t*) APPLICATION_ADDRESS);
    Jump_To_Application();
    }}
```

Build the project.

Open the bootloader Options for Target\Debug tab and open the loadApp.ini file.

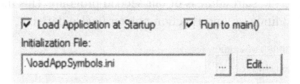

```
Load "C:\\Cortex Microcontroller Tutorial Exercises\\Bootloader\\Blinky\\Flash\
\Blinky.AXF" incremental
```

This script file can be used with the simulator or the hardware debugger. It is used to load the blinky application code as well as the bootloader code. This allows us to debug seamlessly between the two separate programs.

Start the debugger.

Single step the code through the bootloader checking that the correct stack pointer address is loaded into the main stack pointer and that the blinky start address is loaded into the function pointer.

Use the memory window to view the blinky application, which starts at 0×800200, and check that the stack pointer value is loaded into R13 and the reset address is loaded into the function pointer.

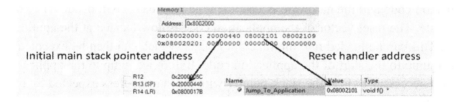

Open the blinky.c file in the blinky project and set a breakpoint on main.

```
35 ⊟int main (void) {
●36      uint32_t ad_avg = 0;
 37      uint16_t ad_val = 0, ad_val_ = 0xFFFF;
```

Run the code.

Now the Cortex processor has left the bootloader program and entered the blinky application program. The startup system code has programmed the vector table offset register so now the hardware vector table matches the blinky software.

Open the Peripherals\Core Peripherals\Nested Vector Interrupt Table.

Now the vector table is at 0x8002000 to match the blinky code.

Open the IRQ.C file and set a breakpoint on the systick interrupt handler.

```
27 void SysTick_Handler (void) {
28     static unsigned long ticks = 0;
29     static unsigned long timetick;
30     static unsigned int  leds = 0x01;
```

Run the code.

When the systick timer raises its interrupt, the handler address will be fetched from the blinky vector table and not from the default address at the start of the memory and the correct systick handler will be executed. Now the blinky program is running happily at its offset address. If you need the application code to call the bootloader, then set a pattern in memory and force a reset within the application code by writing to the SCB application interrupt and reset the control register.

```
Shared_memory = USER_UPDATE_COMMAND
NVIC_SystemReset();
```

When the bootloader restarts part of its startup checks will be to test the shared_memory location and if it is set run the required code.

Power Management

While the Cortex-M0 and Cortex-M0+ are specifically designed for low-power operation the Cortex-M3 and Cortex-M4 still have remarkably low power consumption. While the actual power consumption will depend on the manufacturing process used by the silicon vendor, the figures below give an indication of expected power consumption.

**Table 3.12: Power Consumption Figures by Processor Variant
in 90 nm LP (Low-Power) Process**

Processor	Dynamic Power Consumption (μW/MHz)	Details
Cortex-M0+	11	Excludes debug units
Cortex-M0	16	Excludes debug units
Cortex-M3	32	Excludes MPU and debug units
Cortex-M4	33	Excludes FPU, MPU, and debug units

The Cortex-M processors are capable of entering low-power modes called SLEEP and
DEEPSLEEP. When the processor in placed in SLEEP mode, the main CPU clock signal is
stopped, which halts the Cortex-M processor. The rest of the microcontroller clocks and
peripherals will still be running and can be used to wake up the CPU. The DEEPSLEEP
mode is an extension of the SLEEP mode and its action will depend on the specific
implementation made on the microcontroller. Typically, when the DEEPSLEEP mode is
entered, the processor peripheral clocks will also be halted along with the Cortex-M
processor clock. Other areas of the microcontroller such as the on-chip SRAM and power to
the flash memory may also be switched off depending on the microcontroller configuration.

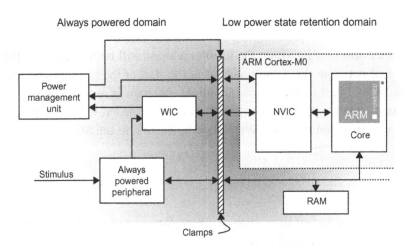

Figure 3.26
The WIC is a small area of gates that do not require a clock source. The WIC can be located on a
different power domain to the Cortex-M processor. This allows all the processor clocks to be
halted. The range of available power modes is defined by the silicon manufacturer.

When a Cortex-M processor has entered a low-power sleep mode, it can be woken up by a
microcontroller peripheral raising an interrupt to the NVIC. However, the NVIC need clock
to operate, so if all the clocks are stopped we need another hardware unit to tell the power
management unit (PMU) to restore the clocks before the NVIC can respond. The Cortex-M
processors can be fitted with an optional unit called the WIC. The WIC handles interrupt

detection when all clocks are stopped and allows all of the Cortex processors to enter low-power modes. The WIC consists of a minimal number of gates and does not need a system clock. The WIC can be placed on a different power domain to the main Cortex-M processor. This allows the microcontroller manufacturers to design a device that can have low-power modes where most of the chip is switched off while keeping key peripherals alive to wake up the processor.

Entering Low-Power Modes

The Thumb-2 instruction set contains two dedicated instructions that will place the Cortex processor into SLEEP or DEEPSLEEP mode.

Table 3.13: Low-Power Entry Instructions

Instruction	Description	CMSIS-Core Intrinsic
WFI	Wait for interrupt	__WFI()
WFE	Wait for event	__WFE

As its name implies the WFI instruction will place the Cortex-M processor in the selected low-power mode. When an interrupt is received from one of the microcontroller peripherals, the processor will exit low-power mode and resume processing the interrupt as normal. The wait for event instruction is also used to enter the low-power modes but has some configuration options as we shall see next.

Configuring the Low-Power Modes

The system control register (SCR) is used to configure the Cortex-M processor's low-power options.

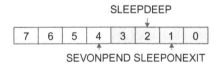

Figure 3.27
The SCR contains the Cortex-M processor's low-power configuration bits.

The Cortex-M processor has two external sleep signals which are connected to the power management system designed by the manufacturer. By default, the SLEEPING signal is activated when the WFI or WFE instructions are executed. If the SLEEPDEEP bit is set, the second sleep signal SLEEPDEEP is activated when the Cortex-M processor enters a sleeping mode. These two sleep signals are used by the microcontroller manufacturer to provide a broader power management scheme for the microcontroller. Setting the sleep on exit bit will force the microcontroller to enter its SLEEP mode when it reaches the end of an interrupt. This allows you to design a system that wakes up in response to an interrupt, runs the required code, and then automatically returns to sleep. In such a system no stack management is required (except in the case of preempted interrupts) during an interrupt entry/exit sequence and no background code will be executed. The WFE instruction places the Cortex-M processor into its sleeping mode and the processor may be woken by an interrupt in the same manner as the WFI instruction. However, the WFE instruction has an internal event latch. If the event latch is set to one of the processor, it will clear the latch but do not enter low-power mode. If the latch is zero, it will enter low-power mode. On a typical microcontroller, the events are the peripheral interrupt signals. So pending interrupts will prevent the processor from sleeping. The SEVONPEND bit is used to change the behavior of the WFE instruction. If this bit is set, the peripheral interrupt lines can be used to wake the processor even if the interrupt is disabled in the NVIC. This allows you to place the processor into its sleep mode, and when a peripheral interrupt occurs the processor will wake and resume execution of the instruction following the WFE instruction rather than jumping to an interrupt routine.

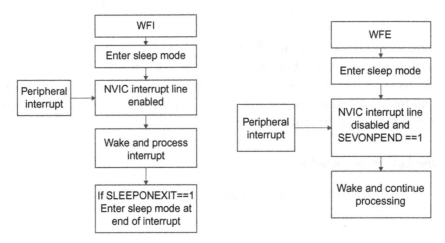

Figure 3.28

The two low-power entry instructions place the Cortex processor into its low-power mode. Both modes use a peripheral interrupt to wake the processor but their wakeup behavior is different.

An interrupt disabled in the NVIC cannot be used to exit a sleep mode entered by the WFI instruction. However, the WFE instruction will respond to an activity on any interrupt line even if it is disabled or temporarily disabled by the processor mask registers (i.e., BASEPRI, PRIMASK, and FAULTMASK).

Exercise: Low-Power Modes

In this exercise, we will use the exception project to experiment with the Cortex-M low-power modes.

Open the project in c:\execises\low power modes.

```
while(1)
{
  SLEEP = 1;
  BACKGROUND = 0;
  __wfe();
  BACKGROUND = 1;
  SLEEP = 0;
}
```

This project uses the systick and ADC interrupts that we saw in the last example. This time we have added an additional SLEEP variable to monitor the processor operating state. The wait for interrupt instruction has been added in the main while loop.

Build the project and start the debugger.

Open the logic analyzer window and start running the code.

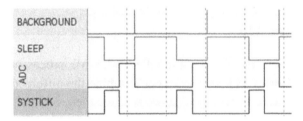

Here, we can see that the background code executes the __WFI() instruction and then goes to sleep until an interrupt is raised. When the interrupts have completed we return to the background code, which will immediately place the processor in its low-power mode.

Exit the debugger and change the code to match the lines below.

```
SCB->SCR = 0x2;
SLEEP = 1;
BACKGROUND = 0;
__wfi();
while(1){
```

Add the code to set bit two of the SCR. This sets the sleep on exit flag, which forces the processor into a low-power mode when it completes an interrupt. Cut the remaining three lines from inside the while loop and paste them into the initializing section of the main() function.

Build the code, restart the debugger, and observe the execution of the interrupts in the logic analyzer window.

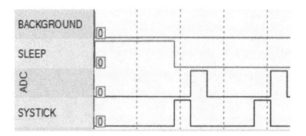

Here, we can see that the interrupts are running but after the initializing code has run the background loop never executes so the background and sleep variables are never updated. In the debugger, you will also be able to see from the coverage monitor that the main() while loop is never executed. This feature allows the processor to wake up, run some critical code, and then sleep with the absolute minimum overhead.

Moving from the Cortex-M3

In this chapter, we have concentrated on learning the Cortex-M3 processor. Now that you are familiar with how the Cortex-M3 works, we can examine the differences between the Cortex-M3 and the other Cortex-M variants. As we will see these are mainly architectural differences, and if you can use the Cortex-M3 you can easily move up to a Cortex-M4 or down to a Cortex-M0(+)-based microcontroller. Increasingly silicon manufacturers are producing families of microcontrollers which have variants using different Cortex-M processors. This allows you to seamlessly switch devices trading off performance against cost.

Cortex-M4

The Cortex-M4 is most easily described as a Cortex-M3 with an additional FPU and DSP instructions. We will look at these features in Chapter 7, but here we will take a tour of the main differences between the Cortex-M3 and Cortex-M4. The Cortex-M4 offers the same processing power of 1.25 DMIPS/MHz as the Cortex-M3 but has a much greater math capability. This is delivered in three ways. The hardware FPU can perform calculations in a little as one cycle compared to the hundreds of cycles the same calculation would take on the Cortex-M3. For integer calculations, the Cortex-M4 has a higher performance MAC that improves on the Cortex-M3 MAC to allow single cycle calculations to be performed on 32-bit wide quantities, which yield a 64-bit result. Finally, the Cortex-M4 adds a group of SIMD instructions that can perform multiple integer calculations in a single cycle.

While the Cortex-M4 has a larger gate count than the Cortex-M3, the FPU contains more gates than the entire Cortex-M0 processor.

Table 3.14: Additional Features in the Cortex-M4

Feature	Comments
FPU	See Chapter 7
DSP, SIMD instructions	
GE field in xPSR	
Extended integer MAC unit	The Cortex-M4 extends the integer MAC to support single cycle execution of 32-bit multiplies which yield a 64-bit result

Cortex-M0

The Cortex-M0 is a reduced version of the Cortex-M3; it is intended for low-cost and low-power microcontrollers. The Cortex-M0 has a processing power of 0.84 DMIPS/MHz. While the Cortex-M0 can run at high clock frequencies it is often designed into low-cost devices with simple memory systems. Hence, a typical Cortex-M0 microcontroller runs with a CPU frequency of 50 DMIPS/MHz, however, its low-power consumption makes it ideal for low-power applications. While it essentially has the same programmer's model as the Cortex-M3, there are some limitations and these are summarized below.

Table 3.15: Features Not Included in the Cortex-M0

Feature	Comments
Three-stage pipeline	Same pipeline stages as the Cortex-M3, but no speculative branch target fetch
von Neumann bus interface	Instruction and data use the same bus port which can create slower performance compared to the Harvard architecture of the Cortex-M3/M4
No conditional if then blocks	
No saturated math instructions	Conditional branches are always used which cause a pipeline flush
SYSTICK timer is optional	However, so far every Cortex-M0 microcontroller had it fitted
No MPU	The MPU is covered in Chapter 5
32 NVIC channels	A limited number of interrupt channels compared to the Cortex-M3/M4, however, in practice this is not a real limitation and the Cortex-M0 is intended for small devices with a limited number of peripherals
Four programmable priority levels	Priority level registers only implemented 2 bits (four levels), and there is no priority group setting
Hard fault exception only	No usage fault, memory management fault, or bus fault exception vectors
16 cycle interrupt latency	Same deterministic interrupt handling as the Cortex-M3 but with four more cycles of overhead

(Continued)

Table 3.15: (Continued)

Feature	Comments
No priority group	Limited to a fixed four levels of preemption
No BASEPRI register	
No Faultmask register	
Reduced debug features	The Cortex-M0 has fewer debug features compared to the Cortex-M3\M4, see Chapter 8 for more details
No exclusive access instructions	The exclusive access instructions are covered in Chapter 5
No reverse bit order of count leading zero instructions	
Reduced number of registers in the SCB	See below
Vector table cannot be relocated	The NVIC does not include the vector table offset register
All code executes at privileged level	The Cortex-M0 does not support the unprivileged operating mode

The SCB contains a reduced number of features compared to the Cortex-M3\M4. Also the systick timer registers have been moved from the NVIC to the SCB.

Table 3.16: Registers in the Cortex-M0 SCB

Register	Size in Words	Description
Systick control and status	1	Enables the timer and its interrupt
Systick reload	1	Holds the 24-bit reload value
Systick current value	1	Holds the current 24-bit timer value
Systick calibration	1	Allows trimming the input clock frequency
CPU ID	1	Hardwired ID and revision numbers from ARM and the silicon manufacturer
Interrupt control and state	1	Provides pend bits for the systick and NMI interrupts and extended interrupt pending\active information
AIRC	1	Contains the same fields as the Cortex-M3 minus the PRIGROUP field
Configuration and control	1	
System handler priority	2	These registers hold the 8-bit priority fields for the configurable processor exceptions

Cortex-M0+

The Cortex-M0+ is the latest Cortex-M processor variant to be released from ARM. It is an enhanced version of the Cortex-M0. As such it boasts lower power consumption figures combined with greater processing power. The Cortex-M0+ also brings the MPU and real-time debug capability to very low-end devices. The Cortex-M0+ also introduces a fast IO port that speeds up access to peripheral registers, typically GPIO ports, to allow fast switching of port pins. As we will see in Chapter 8, the debug system is fitted with a new trace unit called the MTB which allows you to capture a history of executed code with a low-cost development tool.

Table 3.17: Cortex-M0+ Features

Feature	Comments
Code compatible with the Cortex-M0	The Cortex-M0+ is code compatible with the Cortex-M0 and provides higher performance with lower-power consumption
Two-stage pipeline	This reduces the number of flash accesses and hence power consumption
I/O port	The I/O port provides single cycle access to GPIO and peripheral registers
Vector table can be relocated	This supports more sophisticated software designs. Typically a bootloader and separate application
Supports 16-bit flash memory accesses	This allows devices featuring the Cortex-M0+ to use low-cost memory
Code can execute at privileged and unprivileged levels	The Cortex-M0+ has the same operating modes as the Cortex-M3/M4
MPU	The Cortex-M0+ has a similar MPU to the Cortex-M3/M4
MTB	This is a "snapshot" trace unit which can be accessed by low cost debug units

CHAPTER 4

Cortex Microcontroller Software Interface Standard

Like desktop computing the software complexity of embedded applications is increasing exponentially. Now more than ever developers are using third-party code to meet project deadlines. ARM has defined the Cortex Microcontroller Software Interface Standard (CMSIS), which allows easy integration of source code from multiple sources. CMSIS is gaining increasing support through the industry and should be adopted for new projects.

Introduction

The widespread adoption of the Cortex-M processor into general purpose microcontrollers has led to two rising trends within the electronics industry. First of all the same processor is available from a wide range of vendors each with their own family of microcontrollers. In most cases, each vendor creates a range of microcontrollers that span a range of requirements for embedded systems developers. This proliferation of devices means that as a developer you can select a suitable microcontroller from many hundreds of devices while still using the same tools and skills regardless of the silicon vendor. This explosive growth in Cortex-M-based microcontrollers has made the Cortex-M processor the de facto industry standard for 32-bit microcontrollers and there are currently no real challengers.

Figure 4.1
CMSIS compliant software development tools and middleware stacks are allowed to carry the CMSIS logo.

The flip side of the coin is differentiation. It would be possible for a microcontroller vendor to design their own proprietary 32-bit processor. However, this is expensive to do and also requires an ecosystem of affordable tools and software to achieve mass adoption. It is more cost effective to license the Cortex-M processor from ARM and then use their own expertise to create a microcontroller with innovative peripherals. There are now more than 10 silicon vendors shipping Cortex-M-based microcontrollers. While in each device the Cortex-M processor is the same, each silicon manufacturer seeks to offer a unique set of user peripherals for a given range of applications. This can be a microcontroller designed for low-power applications, motor control, communications, or graphics. This way a silicon vendor can offer a microcontroller with a state-of-the-art processor that has wide development tools support while at the same time using their skill and knowledge to develop a microcontroller featuring an innovative set of peripherals.

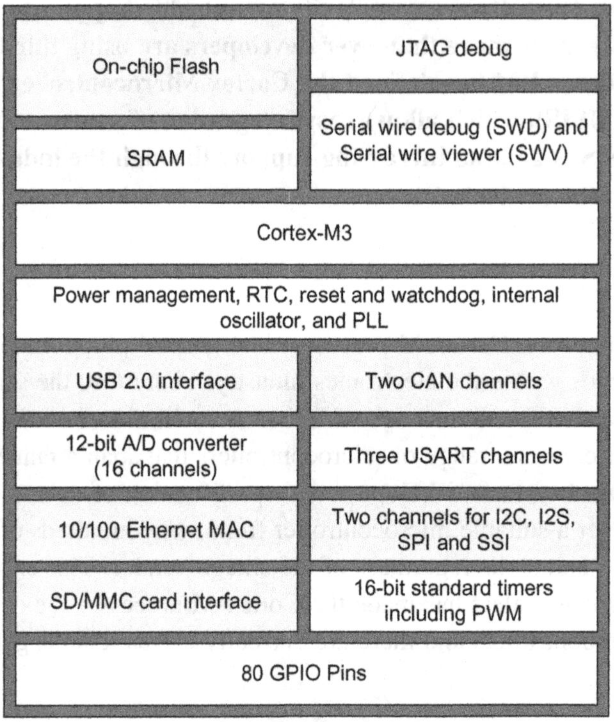

On-chip Flash	JTAG debug
SRAM	Serial wire debug (SWD) and Serial wire viewer (SWV)
Cortex-M3	
Power management, RTC, reset and watchdog, internal oscillator, and PLL	
USB 2.0 interface	Two CAN channels
12-bit A/D converter (16 channels)	Three USART channels
10/100 Ethernet MAC	Two channels for I2C, I2S, SPI and SSI
SD/MMC card interface	16-bit standard timers including PWM
80 GPIO Pins	

Figure 4.2
Cortex-based microcontrollers can have a number of complex peripherals on a single chip. To make these work you will need to use some form of third-party code. CMSIS is intended to allow stacks from different sources to integrate together easily.

These twin factors have led to a vast "cloud" of standard microcontrollers with increasingly complex peripherals as well as typical microcontroller peripherals such as USART, I2C, ADC, and DAC. A modern high-end microcontroller could well have a Host\Device USB controller, Ethernet MAC, SDIO controller, and LCD interface. The software to drive any of these peripherals is effectively a project in itself, so gone are the days of a developer using an 8\16-bit microcontroller and writing all of the application code from the reset vector. To release any kind of sophisticated product it is almost certain that you will be using some form of third-party code in order to meet project deadlines. The third-party code may take the form of example code, an open source or commercial stack or a library provided by the silicon vendor. Both of these trends have created a need to make C-level code more portable between different development tools and different microcontrollers. There is also a need to be able to easily integrate code taken from a variety of sources into a project.

In order to address these issues, a consortium of silicon vendors and tools vendors has developed the CMSIS (seeMsys) for short.

CMSIS Specifications

The main aim of CMSIS is to improve software portability and reusability across different microcontrollers and toolchains. This allows software from different sources to integrate seamlessly together. Once learned, CMSIS helps to speed up software development through the use of standardized software functions.

At this point it is worth being clear about exactly what CMSIS is. CMSIS consists of five interlocking specifications that support code development across all Cortex-M-based microcontrollers. The four specifications are as follows: CMSIS core, CMSIS RTOS, CMSIS DSP, CMSIS SVD, and CMSIS DAP.

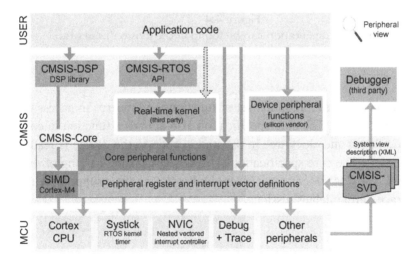

Figure 4.3
CMSIS consists of a several separate specifications (CORE, DSP, RTOS, SVD, and DAP) that make source code more portable between tools and devices.

It is also worth being clear what CMSIS is not. CMSIS is not a complex abstraction layer that forces you to use a complex and bulky library. CMSIS does not attempt to "dumb down" peripherals by providing standard profiles that make different manufacturers' peripherals work the same way. Rather, the CMSIS core specification takes a very small amount of resources (about 1 k of code and just 4 bytes of RAM) and just standardizes the way you access the Cortex-M processor and microcontroller registers. Furthermore, CMSIS does not really affect the way you develop code or force you to adopt a particular methodology. It simply provides a framework that helps you to integrate third-party code and reuse the code on future projects. Each of the CMSIS specifications are not that complicated and can be learned easily.

The full CMSIS specifications can be downloaded from the URL www.onarm.com. Each of the CMSIS specifications are integrated into the MDK-ARM toolchain and the CMSIS documentation is available from the online help.

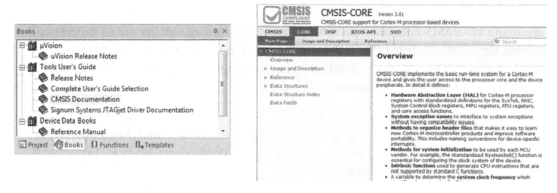

Figure 4.4
The CMSIS documentation can be found in the μVision books tab.

CMSIS Core

The core specification provides a minimal set of functions and macros to access the key Cortex-M processor registers. The core specification also defines a function to configure the microcontroller oscillators and clock tree in the startup code so the device is ready for use when you reach main(). The core specification also standardizes the naming of the device peripheral registers. The CMSIS core specification also standardizes support for the instrumentation trace during debug sessions.

CMSIS RTOS

The CMSIS RTOS specification provides a standard API for an RTOS. This is in effect a set of wrapper functions that translate the CMSIS RTOS API to the API of the specific RTOS that you are using. We will look at the use of an RTOS in general and the CMSIS RTOS API in Chapter 6. The Keil RTX RTOS was the first RTOS to support the CMSIS RTOS API and it has been released as an open source reference implementation. RTX can be compiled with both the GCC and IAR compilers. It is licensed with a three-clause Berkeley Software Distribution (BSD) license that allows its unrestricted use in commercial and noncommercial applications.

CMSIS DSP

As we have seen in Chapter 3, the Cortex-M4 is a "digital signal controller" with a number of enhancements to support DSP algorithms. Developing a real-time DSP system is best described as a "nontrivial pastime" and can be quite daunting for all but the simplest systems. To help mere mortals include DSP algorithms in Cortex-M4 and Cortex-M3 projects. CMSIS includes a DSP library that provides over 60 of the most commonly used DSP mathematical functions. These functions are optimized to run on the Cortex-M4 but can also be compiled to run on the Cortex-M3. We will take a look at using this library in Chapter 7.

CMSIS SVD and DAP

One of the key problems for tools vendors is to provide debug support for new devices as soon as they are released. The debugger must provide peripheral view windows that show the developer the current state of the microcontroller peripheral registers. With the growth in both the numbers of Cortex-M vendors and also the rising number and complexity of on-chip peripherals it is becoming all but impossible for any given tools vendor to maintain support for all possible microcontrollers. To overcome this hurdle, the CMSIS debug specification defines a "system viewer description"(SVD) file. This file is provided and maintained by the silicon vendor and contains a complete description of the microcontroller peripheral registers in an XML format. This file is then imported by the development tool, which uses it to automatically construct the peripheral debug windows for the microcontroller. This approach allows full debugger support to be available as new microcontrollers are released.

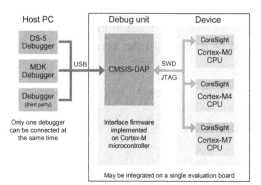

Figure 4.5
CMSIS DAP allows for interoperability between different vendors, software, and hardware debuggers.

The CMSIS DAP specification defines the interface protocol for a hardware debug unit that sits between the host PC and the debug access port (DAP) of the microcontroller. This allows any software toolchain that supports CMSIS DAP to connect to any hardware debug unit that also supports CMSIS DAP. There are an increasing number of very-low-cost evaluation boards that contain an integral debugger. Often this debugger supports a selected toolchain; with CMSIS DAP such a board could be used with any compliant development tool.

Foundations of CMSIS

The CMSIS core specification provides a standard set of low-level functions, macros, and peripheral register definitions that allow your application code to easily access the Cortex-M processor and microcontroller peripheral registers. This framework needs to be added to your code at the start of a project. This is actually very easy to do as the CMSIS core functions are very much a part of the compiler toolchain.

Coding Rules

While CMSIS is important for providing a standardized software interface for all Cortex-M microcontrollers it is also interesting for embedded developers because it defines a consistent set of C coding rules. When applied, these coding rules generate clear unambiguous C code. This approach is worth studying as it embodies many of the best practices that should be adopted when writing the C source code for your own application software.

MISRA C

The main backbone of the CMSIS coding rules is a set of coding guidelines called MISRA C published by MIRA. MIRA stands for "Motor Industry Research Agency" and is located near Rugby in England. It is responsible for many of the industry standards used by the UK motor industry. In 1998, its software division the Motor Industry Software Research Agency released the first version of its coding rules formally called "MISRA guidelines for the use of C in vehicle electronics."

Figure 4.6
The CMSIS source code has been developed using MISRA C as a coding standard.

The original MISRA C specification contained 127 rules which attempted to prevent common coding mistakes and resolve gray areas of the ANSI C specification when applied to embedded systems. Although originally intended for the automotive industry MISRA C has found acceptance in the wider embedded systems community. In 2004, a revised edition of MISRA C was released with the title "MISRA C coding guidelines for the use of C in safety systems." This change in the title reflects the growing adoption of MISRA C as a coding standard for general embedded systems. One of the other key attractions of MISRA C is that it was written by engineers and not by computer scientists. This has resulted in a clear, compact, and easy to understand set of rules. Each rule is clearly explained with examples of good coding practice. This means that the entire coding standard is contained in a book of just 106 pages which can easily be read in an evening. A typical example of a MISRA C rule is shown below.

Rule 13.6(required) Numeric variables being used within a for loop for iteration counting shall not be modified in the body of the loop.

Loop counters shall not be modified in the body of the loop. However, other loop control variables representing logical values may be modified in the loop. For example, a flag to indicate that something has been completed which is then tested in the for statement.

```
Flag = 1;
For ((I = 0;(i < 5) && (flag == 1);i++)
{
/*.......*/
Flag = 0;   /* Compliant – allows early termination of the loop */
i = i + 3;    /*Not Compliant – altering the loop counter */
}
```

Where possible, the MISRA C rules have been designed so that they can be statically checked either manually or by a dedicated tool. The MISRA C standard is not an open standard and is published in paper and electronic form on the MIRA Web site. Full details of how to obtain the MISRA standard are available in Appendix A.

In addition to the MISRA C guidelines, CMSIS enforces some additional coding rules. To prevent any ambiguity in the compiler implementation of standard C types CMSIS uses the data types defined in the ANSI C header file stdint.h.

The typedefs ensure that the expected data size is mapped to the correct ANSI type for a given compiler. Using typedefs like this is a good practice, as it avoids any ambiguity about the underlying variable size, which may vary between compilers particularly if you are migrating code between different processor architectures and compiler tools.

CMSIS also specifies IO type qualifiers for accessing peripheral variables. These are typedefs that make clear the type of access each peripheral register has.

Table 4.1: CMSIS Variable Types

Standard ANSI C Type	MISRA C Type
Signed char	int8_t
Signed short	int16_t
Signed int	int32_t
Signed __int64	int64_t
Unsigned char	uint8_t
Unsigned short	uint16_t
Unsigned int	uint32_t
Unsigned __int64	uint64_t

While this does not provide any extra functionality for your code it provides a common mechanism that can be used by static checking tools to ensure that the correct access is made to each peripheral register.

Much of the CMSIS documentation is autogenerated using a tool called Doxegen. This is a free download released under a GNU Public (GPL) license. While Doxegen cannot

Table 4.2: CMSIS IO Qualifiers

MISRA C IO Qualifier	ANSI C Type	Description
#define __I	Volatile const	Read only
#define __O	Volatile	Write only
#define __IO	Volatile	Read and write

actually write the documentation for you it does do much of the dull boring stuff for you (leaving you to do the exiting documentation work). Doxegen works by analyzing your source code and extracting declarations and specific source code comments to build up a comprehensive "object dictionary" for your project. The default output format for Doxegen is a browsable HTML but this can be converted to other formats if desired. The CMSIS source code comments contain specific tags prefixed by the @ symbol, for example, @brief. These tags are used by Doxegen to annotate descriptions of the CMSIS functions.

```
/**
* @brief Enable Interrupt in NVIC Interrupt Controller
* @param IRQn interrupt number that specifies the interrupt
* @return none.
* Enable the specified interrupt in the NVIC Interrupt Controller.
* Other settings of the interrupt such as priority are not affected.
*/
```

When the Doxegen tool is run it analyzes your source code and generates a report containing a dictionary of your functions and variables based on the comments and source code declarations.

CMSIS Core Structure

The CMSIS core functions can be included in your project through the addition of three files. These include the default startup code with the CMSIS standard vector table. The second file is the system_<device>.c file, which contains the necessary code to initialize the microcontroller system peripherals. The final file is the device include file, which imports the CMSIS header files that contain the CMSIS core functions and macros.

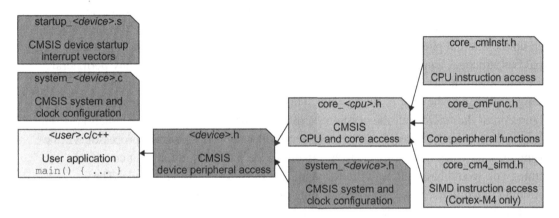

Figure 4.7
The CMSIS core standard consists of the device startup, system C code, and a device header. The device header defines the device peripheral registers and pulls in the CMSIS header files. The CMSIS header files contain all of the CMSIS core functions.

Startup Code

The startup code provides the reset vector, initial stack pointer value, and a symbol for each of the interrupt vectors.

```
__Vectors  DCD  __initial_sp    ; Top of Stack
DCD Reset_Handler               ; Reset Handler
DCD NMI_Handler                 ; NMI Handler
DCD HardFault_Handler           ; Hard Fault Handler
DCD MemManage_Handler           ; MPU Fault Handler
```

When the processor starts, it will initialize the MSP by loading the value stored in the first 4 bytes of the vector table. Then it will jump to the reset handler.

```
Reset_Handler  PROC
  EXPORT  Reset_Handler    [WEAK]
```

```
IMPORT __main
IMPORT SystemInit
  LDR  R0, = SystemInit
  BLX  R0
  LDR  R0, = __main
  BX  R0
  ENDP
```

System Code

The reset handler calls the SystemInit() function, which is located in the CMSIS system_ < device>.c file. This code is delivered by the silicon manufacturer and it provides all the necessary code to configure the microcontroller out of reset. Typically this includes setting up the internal phase-locked loops, configuring the microcontroller clock tree and internal bus structure, enabling the external bus if required, and switching on any peripherals held in low-power mode. The configuration of the initializing functions is controlled by a set of #defines located at the start of the module. This allows you to customize the basic configuration of the microcontroller system peripherals. Since the SystemInit() function is run when the microcontroller leaves reset the microcontroller and the Cortex-M processor will be in a running state when the program reaches main. In the past, this initializing code was something you would have had to write for yourself or crib from example code. The SystemInit() function does save you a lot of time and effort. The SystemInit() function also sets the CMSIS global variable SystemCoreClock to the CPU frequency. This variable can then be used by the application code as a reference value when configuring the microcontroller peripherals. In addition to the SystemInit() function the CMSIS system file contains an additional function to update the SystemCoreClock variable if the CPU clock frequency is changed on the fly. The function SystemCoreClockUpdate() is a void function that must be called if the CPU clock frequency is changed. This function is tailored to each microcontroller and will evaluate the clock tree registers to calculate the new CPU operating frequency and change the SystemCoreClock variable accordingly.

Once the SystemInit() function has run and we reach the application code we will need to access the CMSIS core functions. This framework is added to the application modules through the microcontroller-specific header file.

Device Header File

The header file first defines all of the microcontroller special function registers in a CMSIS standard format.

A typedef structure is defined for each group of special function registers on the supported microcontroller. In the code below, a general GPIO typedef is declared for the group of

GPIO reregisters. This is a standard typedef but we are using the IO qualifiers to designate the type of access granted to a given register.

```
typedef struct
{
__IO uint32_t MODER;   /*!< GPIO port mode register,        Address offset: 0x00 */
__IO uint32_t OTYPER;  /*!< GPIO port output type register,    Address offset: 0x04 */
__IO uint32_t OSPEEDR;  /*!< GPIO port output speed register,    Address offset: 0x08 */
__IO uint32_t PUPDR;  /*!< GPIO port pull-up/pull-down register,  Address offset: 0x0C */
__IO uint32_t IDR;  /*!< GPIO port input data register,    Address offset: 0x10 */
__IO uint32_t ODR;  /*!< GPIO port output data register,    Address offset: 0x14 */
__IO uint16_t BSRRL;  /*!< GPIO port bit set/reset low register,  Address offset: 0x18 */
__IO uint16_t BSRRH;  /*!< GPIO port bit set/reset high register,  Address offset: 0x1A */
__IO uint32_t LCKR;  /*!< GPIO port configuration lock register,  Address offset: 0x1C */
__IO uint32_t AFR[2];  /*!< GPIO alternate function registers,  Address offset:
                                                 0x24-0x28 */
} GPIO_TypeDef;
```

Next #defines are used to lay out the microcontroller memory map. First, the base address of the peripheral special function registers is declared and then offset addresses to each of the peripheral busses and finally an offset to the base address of each GPIO port.

```
#define PERIPH_BASE    ((uint32_t)0x40000000)
#define APB1PERIPH_BASE    PERIPH_BASE
#define GPIOA_BASE  (AHB1PERIPH_BASE + 0x0000)
#define GPIOB_BASE  (AHB1PERIPH_BASE + 0x0400)
#define GPIOC_BASE  (AHB1PERIPH_BASE + 0x0800)
#define GPIOD_BASE  (AHB1PERIPH_BASE + 0x0C00)
```

Then the register symbols for each GPIO port can be declared.

```
#define GPIOA   ((GPIO_TypeDef *)GPIOA_BASE)
#define GPIOB  ((GPIO_TypeDef *) GPIOB_BASE)
#define GPIOC  ((GPIO_TypeDef *) GPIOC_BASE)
#define GPIOD  ((GPIO_TypeDef *) GPIOD_BASE)
```

Then in the application code we can program the peripheral special function registers by accessing the structure elements.

```
void LED_Init (void) {
RCC->AHB1ENR |= ((1UL << 3) ); /* Enable GPIOD clock    */
GPIOD->MODER &= ~((3UL << 2*12) |
    (3UL << 2*13) |
    (3UL << 2*14) |
```

```
        (3UL << 2*15) ); /* PD.12..15 is output    */
GPIOD->MODER |= ((1UL << 2*12)|
    (1UL << 2*13)|
    (1UL << 2*14)|
    (1UL << 2*15) );
```

The microcontroller include file provides similar definitions for all of the on-chip peripheral special function registers. These definitions are created and maintained by the silicon manufacturer and as they do not use any non-ANSI keywords the include file may be used with any C compiler. This means that any peripheral driver code written to the CMSIS specification is fully portable between CMSIS compliant tools.

The microcontroller include file also provides definitions of the interrupt channel number for each peripheral interrupt source.

```
WWDG_IRQn        = 0,  /*!< Window WatchDog interrupt    */
PVD_IRQn         = 1,  /*!< PVD through EXTI line detection interrupt    */
TAMP_STAMP_IRQn  = 2,  /*!< Tamper and TimeStamp interrupts through the EXTI line    */
RTC_WKUP_IRQn    = 3,  /*!< RTC Wakeup interrupt through the EXTI line    */
FLASH_IRQn       = 4,  /*!< FLASH global interrupt    */
RCC_IRQn         = 5,  /*!< RCC global interrupt    */
EXTI0_IRQn       = 6,  /*!< EXTI Line0 interrupt    */
EXTI1_IRQn       = 7,  /*!< EXTI Line1 interrupt    */
EXTI2_IRQn       = 8,  /*!< EXTI Line2 interrupt    */
EXTI3_IRQn       = 9,  /*!< EXTI Line3 interrupt    */
EXTI4_IRQn       = 10, /*!< EXTI Line4 interrupt
```

In addition to the register and interrupt definitions, the silicon manufacturer may also provide a library of peripheral driver functions. Again as this code is written to the CMSIS standard it will compile with any suitable development tool. Often these libraries are very useful for getting a project to work quickly and minimizing the amount of time you have to spend writing low-level code. However, they are often very general libraries that do not yield the most optimized code. So, if you need to get the maximum performance/minimal code size, you will need to rewrite the driver functions to suit your specific application. The microcontroller include file also imports up to five further include files. These include stdint.h, a CMSIS core file for the Cortex processor you are using. A header file System_<device>.h is also included to give access to the functions in the system file. The CMSIS instruction intrinsic and helper functions are contained in two further files, core_cminstr.h and core_cmfunc.h. If you are using the Cortex-M4 an additional file, core_CM4_simd.h, is added to provide support for the Cortex-M4 SIMD instructions.

As discussed earlier, the stdint.h file provides the MISRA C types that are used in the CMSIS definitions and should be used through your application code.

CMSIS Core Header Files

The device file imports the Cortex-M processor which are held in their own include files. There are a small number of defines that are set up for a given device. These can be found in the main processor include file.

The processor include file also imports the CMSIS header files, which contain the CMSIS core helper functions. The helper functions are split into the groups shown below.

The NVIC group provides all the functions necessary to configure the Cortex-M interrupts and exceptions. A similar function is provided to configure the systick timer and interrupt. The CPU register group allows you to easily read and write to the CPU registers using the MRS and MSR instructions. Any instructions that are not reachable by the C language are

Table 4.3: CMSIS Configuration Values

__CMx_REV	Core revision number
__NVIC_PRIO_BITS	Number of priority bits implemented in the NVIC priority registers
__MPU_PRESENT	Defines if an MPU is present (see Chapter 5)
__FPU_PRESENT	Defines if an FPU is present (see Chapter 7)
__Vendor_SysTickConfig	Defines if there is a vendor-specific systick configuration

provided by dedicated intrinsic functions and are contained in the CPU instructions group. An extended set of intrinsics are also provided for the Cortex-M4 to access the SIMD instructions. Finally, some standard functions are provided to access the debug instrumentation trace.

Interrupts and Exceptions

Management of the NVIC registers may be done by the functions provided in the interrupt and exception group. These functions allow you to setup an NVIC interrupt channel and manage its priority as well as interrogate the NVIC registers during runtime.

A configuration function is also provided for the systick timer.

So, for example, to configure an external interrupt line we first need to find the name for the external interrupt vector used in the startup code of the vector table.

Table 4.4: CMSIS Function Groups

CMSIS Core Function Groups
NVIC access functions
Systick configuration
CPU register access
CPU instruction intrinsics
Cortex-M4 SIMD intrinsics
ITM debug functions

Table 4.5: CMSIS Interrupt and Exception Group

CMSIS Function	Description
NVIC_SetPriorityGrouping	Set the priority grouping (Not for Cortex-M0)
NVIC_GetPriorityGrouping	Read the priority grouping (Not for Cortex M0)
NVIC_EnableIRQ	Enable a peripheral interrupt channel
NVIC_DisableIRQ	Disable a peripheral interrupt channel
NVIC_GetPendingIRQ	Read the pending status of an interrupt channel
NVIC_SetPendingIRQ	Set the pending status of an interrupt channel
NVIC_ClearPendingIRQ	Clear the pending status of an interrupt channel
NVIC_GetActive	Get the active status of an interrupt channel (Not for Cortex-M0)
NVIC_SetPriority	Set the active status of an interrupt channel
NVIC_GetPriority	Get the priority of an interrupt channel
NVIC_EncodePriority	Encode the priority group (Not for Cortex-M0)
NVIC_DecodePriority	Decode the priority group (Not for Cortex-M0)
NVIC_SystemReset	Force a system reset

Table 4.6: CMSIS Systick Function

CMSIS Function	Description
SysTick_Config	Configures the timer and enables the interrupt

```
    DCD FLASH_IRQHandler          ; FLASH
DCD   RCC_IRQHandler              ; RCC
DCD   EXTI0_IRQHandler            ; EXTI Line0
DCD   EXTI1_IRQHandler            ; EXTI Line1
DCD   EXTI2_IRQHandler            ; EXTI Line2
DCD   EXTI3_IRQHandler            ; EXTI Line3
DCD   EXTI4_IRQHandler            ; EXTI Line4
DCD   DMA1_Stream0_IRQHandler     ; DMA1 Stream 0
```

So for external interrupt line 0 we simply need to create a void function duplicating the name used in the vector table.

```
void EXTIO_IRQHandler (void);
```

This now becomes our interrupt service routine. In addition, we must configure the microcontroller peripheral and NVIC to enable the interrupt channel. In the case of an external interrupt line, the following code will setup Port A pin 0 to generate an interrupt to the NVIC on a falling edge.

```
AFIO->EXTICR[0]  &= ~AFIO_EXTICR1_EXTIO;      /* clear used pin */
AFIO->EXTICR[0]  |= AFIO_EXTICR1_EXTIO_PA;    /* set PA.0 to use */
EXTI->IMR        |= EXTI_IMR_MRO;             /* unmask interrupt */
EXTI->EMR        &= ~EXTI_EMR_MRO;            /* no event */
EXTI->RTSR       &= ~EXTI_RTSR_TRO;           /* no rising edge trigger */
EXTI->FTSR       |= EXTI_FTSR_TRO;            /* set falling edge trigger */
```

Next we can use the CMSIS functions to enable the interrupt channel.

```
NVIC_EnableIRQ( EXTIO_IRQn );
```

Here we are using the defined enumerated type for the interrupt channel number. This is declared in the microcontroller header file. Once you get a bit familiar with the CMSIS core functions, it becomes easy to intuitively work out the name rather than having to look it up or look up the NVIC channel number.

We can also add a second interrupt source by using the systick configuration function, which is the only function in the systick group.

```
uint32_t SysTick_Config(uint32_t ticks)
```

This function configures the countdown value of the systick timer and enables its interrupt so an exception will be raised when its count reaches zero. Since the SystemInit() function sets the global variable SystemCoreClock with the CPU frequency that is also used by the systick timer we can easily setup the systick timer to generate a desired periodic interrupt. So a 1 ms interrupt can be generated as follows:

```
Systick_Config(SystemCoreClock/1000);
```

Again we can look up the exception handler from the vector table

```
    DCD  0                 ; Reserved
    DCD  PendSV_Handler    ; PendSV Handler
    DCD  SysTick_Handler   ; SysTick Handler
```

and create a matching C function:

```
void SysTick_Handler (void);
```

Now that we have two interrupt sources we can use other CMSIS interrupt and exception functions to manage the priority levels. The number of priority levels will depend on how many priority bits have been implemented by the silicon manufacturer. For all of the Cortex-M processors we can use a simple "flat" priority scheme where zero is the highest priority. The priority level is set by

```
NVIC_SetPriority(IRQn_Type IRQn, uint32_t priority);
```

The set priority function is a bit more intelligent than a simple macro. It uses the IRQn NVIC channel number to differentiate between user peripherals and the Cortex processor exceptions. This allows it to program either the system handler priority registers in the SCB or the interrupt priority registers in the NVIC itself. The set priority function also uses the NVIC_PRIO_BITS definition to shift the priority value into the active priority bits that have been implemented by the device manufacturer.

```
_STATIC_INLINE void NVIC_SetPriority(IRQn_Type IRQn, uint32_t priority)
{
  if(IRQn<0) {
    SCB->SHP[((uint32_t)(IRQn) & 0xF)-4] = ((priority << (8 - __NVIC_PRIO_BITS)) &
0xff); } /* set Priority for Cortex M System Interrupts */
  else {
    NVIC->IP[(uint32_t)(IRQn)] = ((priority << (8 - __NVIC_PRIO_BITS)) & 0xff);  }
                                  /* set Priority for device specific Interrupts */
}
```

However, for the Cortex-M3 and Cortex-M4 we have the option to set priority groups and subgroups as discussed in Chapter 3. Depending on the number of priority bits defined by the manufacturer, we can configure priority groups and subgroups.

```
NVIC_SetPriorityGrouping();
```

To set the NVIC priority grouping you must write to the AIRC register. As discussed in Chapter 3, this register is protected by its VECTKEY field. In order to update this register, you must write 0x5FA to the VECTKEY field. The SetPriorityGrouping function provides all the necessary code to do this.

```
__STATIC_INLINE void NVIC_SetPriorityGrouping(uint32_t PriorityGroup)
{
uint32_t reg_value;
uint32_t PriorityGroupTmp = (PriorityGroup & (uint32_t)0x07);    /* only values 0..7 are
                                                    used   */
reg_value = SCB->AIRCR;    /* read old register configuration  */
reg_value &= ~(SCB_AIRCR_VECTKEY_Msk | SCB_AIRCR_PRIGROUP_Msk); /* clear bits to
                                                    change   */
```

```
reg_value = (reg_value | ((uint32_t)0x5FA << SCB_AIRCR_VECTKEY_Pos) | /* insert write
                                                        key and priorty group */
  (PriorityGroupTmp << 8));
SCB->AIRCR = reg_value;
}
```

The interrupt and exception group also provides a system reset function that will generate a hard reset of the whole microcontroller.

```
NVIC_SystemReset(void);
```

This function writes to bit 2 of the "Application Interrupt Reset Control" register. This strobes a logic line out of the Cortex-M core to the microcontroller reset circuitry which resets the microcontroller peripherals and the Cortex-M core. However, you should be a little careful here as the implementation of this feature is down to the microcontroller manufacturer and may not be fully implemented. So if you are going to use this feature you need to test it first. Bit zero of the same register will do a warm reset of the Cortex-M core, that is, force a reset of the Cortex-M processor but leave the microcontroller registers configured. This feature is normally used by debug tools.

Exercise: CMSIS and User Code Comparison

In this exercise, we will revisit the multiple interrupts example and examine a rewrite of the code using the CMSIS core functions.

Open the project in c:\exercises\CMSIS core multiple interrupt and c:\exercises\ multiple interrupt.

Open main.c in both projects and compare the initializing code.

The systick timer and ADC interrupts can be initialized with the following CMSIS functions.

```
SysTick_Config(SystemCoreClock / 100);
NVIC_EnableIRQ (ADC1_2_IRQn);
NVIC_SetPriorityGrouping (5);
NVIC_SetPriority (SysTick_IRQn,4);
NVIC_SetPriority (ADC1_2_IRQn,4);
```

Or you can use the equivalent non-CMSIS code.

```
SysTick->VAL = 0x9000;            //Start value for the sysTick counter
SysTick->LOAD = 0x9000;           //Reload value
SysTick->CTRL = SYSTICK_INTERRUPT_ENABLE
```

```
     |SYSTICK_COUNT_ENABLE;          //Start and enable interrupt
  NVIC->ISER[0] = (1UL << 18);      /* enable ADC Interrupt  */
  NVIC->IP[18] = (2<<6 | 2<<4);
  SCB->SHP[11] = (1<<6 | 3<<4);
  Temp = SCB->AIRC;
  Temp &= ~0x
  Temp = Temp|(0xAF0 << )|(0x05 << );
```

Although both blocks of code achieve the same thing, the CMSIS version is much more readable and far less prone to coding mistakes.

Build both projects and compare the size of the code produced.

The CMSIS functions introduce a small overhead but this is an acceptable trade-off against ease of use and maintainability.

CMSIS Core Register Access

The next group of CMSIS functions gives you direct access to the processor core registers.

These functions provide you with the ability to globally control the NVIC interrupts and set the configuration of the Cortex-M processor into its more advanced operating mode. First, we can globally enable and disable the microcontroller interrupts with the following functions.

```
__set_PRIMASK(void);
__set_FAULTMASK (void);
__enable-IRQ
__enable_Fault-irq
__set_BASEPRI()
```

Table 4.7: CMSIS CPU Register Functions

Core Function	Description
__get_Control	Read the control register
__set_Control	Write to the control register
__get_IPSR	Read the IPSR register
__get_APSR	Read the APSR register
__get_xPSR	Read the xPSR register
__get_PSP	Read the process stack pointer
__set_PSP	Write to the process stack pointer
__get_MSP	Read the main stack pointer
__set_MSP	Write to the main stack pointer
__get_PRIMASK	Read the PRIMASK

(Continued)

Table 4.7: (Continued)

Core Function	Description
__set_PRIMASK	Write to the PRIMASK
__get_BASEPRI	Read the BASEPRI register
__set_BASEPRI	Write to the BASEPRI register
__get_FAULTMASK	Read the FAULTMASK
__set_FAULTMASK	Write to the FAULTMASK
__get_FPSCR	Read the FPSCR
__set_FPSCR	Write to the FPSCR
__enable_irq	Enable interrupts and configurable fault exceptions
__disable_irq	Disable interrupts and configurable fault exceptions
__enable_fault_irq	Enable interrupts and all fault handlers
__disable_fault_irq	Disable interrupts and all fault handlers

While all of these functions are enabling and disabling interrupt sources they all have slightly different effects. The __set_PRIMASK() function and the enable_IRQ\Disable_IRQ functions have the same effect in that they set and clear the PRIMASK bit, which enables and disables all interrupt sources except the hard fault handler and the nonmaskable interrupt. The __set_FAULTMASK() function can be used to disable all interrupts except the nonmaskable interrupt. We will see later how this can be useful when we want to bypass the MPU. Finally, the __set_BASEPRI() function sets the minimum active priority level for user peripheral interrupts. When the base priority register is set to a nonzero level the interrupt at the same priority level or lower will be disabled.

These functions allow you to read the PSR and its aliases. You can also access the control register to enable the advanced operating modes of the Cortex-M processor as well as explicitly setting the stack pointer values. A dedicated function is also provided to access the floating point status control (FPSC) register, if you are using the Cortex-M4. We will take a closer look at the more advanced operating modes of the Cortex-M processor in Chapter 5.

CMSIS Core CPU Intrinsic Instructions

The CMSIS core header also provides two groups of standardized intrinsic functions. The first group is common to all Cortex-M processors and the second group provides standard intrinsic for the Cortex-M4 SIMD instructions.

The CPU intrinsics provide direct access to Cortex-M processor instructions that are not directly reachable from the C language. While many of their functions can be achieved by using multiple instructions generated by high-level C code you can optimize your code by the judicious use of these instructions.

With the CPU intrinsics we can enter the low-power modes using the __WFI() and _WFE() instructions. The CPU intrinsics also provide access to the saturated math instructions that we met in Chapter 3. The intrinsic functions also give access to the execution of barrier instructions that ensure completion of a data write or instruction for execution before continuing with the next instruction. The next group of instruction intrinsics is used to guarantee exclusive access to a memory region by one region of code. We will take a look at these in Chapter 5. The remainder of the CPU intrinsics

Table 4.8: CMSIS Instruction Intrinsics

CMSIS Function	Description	More Information
__NOP	No operation	
__WFI	Wait for interrupt	See Chapter 3
__WFE	Wait for event	See Chapter 3
__SEV	Send event	
__ISB	Instruction synchronization barrier	See Chapter 3
__DSB	Data synchronization barrier	See Chapter 3
__DMD	Data memory synchronization barrier	See Chapter 3
__REV	Reverse byte order (32 bit)	See Chapter 4 for rotation instructions
__REV16	Reverse byte order (16 bit)	
__REVSH	Reverse byte order, signed short	
__RBIT	Reverse bit order (not for Cortex-M0)	
__ROR	Rotate right by n bits	
__LDREXB	Load exclusive (8 bits)	See Chapter 5 for exclusive access instructions
__LDREXH	Load exclusive (16 bits)	
__LDREXW	Load exclusive (32 bits)	
__STREXB	Store exclusive (8 bits)	
__STREXH	Store exclusive (16 bits)	
__STREXW	Store exclusive (32 bits)	
__CLREX	Remove exclusive lock	
__SSAT	Signed saturate	See Chapter 3
__USAT	Unsigned saturate	
__CLZ	Count leading zeros	See Chapter 4

support single-cycle data manipulation functions such as the rotate and RBIT instructions.

Exercise: Intrinsic Bit Manipulation

In this exercise we will look at the data manipulation intrinsic supported in CMSIS.

Open the exercise in c:\exercises\CMSIS core intrinsic.

The exercise declares an input variable and a group of output variables and then uses each of the intrinsic data manipulation functions.

```
outputREV    = __REV(input);
outputREV16  = __REV16(input);
outputREVSH  = __REVSH(input);
outputRBIT   = __RBIT(input);
outputROR    = __ROR(input,8);
outputCLZ    = __CLZ(input);
```

Build the project and start the debugger.

Add the input and each of the output variables to the watch window.

Step through the code and count the cycles taken for each function.

While each intrinsic instruction takes a single cycle some surrounding instructions are required so the intrinsic functions take between 9 and 18 cycles.

Examine the values in the output variables to familiarize yourself with the action of each intrinsic.

Name	Value	Type
input	0x00112233	unsigned int
outputREV	0x33221100	unsigned int
outputREV16	0x11003322	unsigned int
outputREVSH	0x00003322	int
outputRBIT	0xCC448800	unsigned int
outputROR	0x33001122	unsigned int
outputCLZ	0x0000000B	unsigned int

Consider how you would code each intrinsic in pure C.

CMSIS SIMD Intrinsics

The final group of CMSIS intrinsics provide direct access to the Cortex-M4 SIMD instructions.

The SIMD instructions provide simultaneous calculations for two 16-bit operations or four 8-bit operations. This greatly enhances any form of repetitive calculation over a data set, as in a digital filter. We will take a close look at these instructions in Chapter 7.

CMSIS Core Debug Functions

The final group of CMSIS core functions provides enhanced debug support through the CoreSight instrumentation trace. The CMSIS standard has two dedicated debug specifications, CMSIS SVD and CMSIS DAP, which we will look at in Chapter 8.

However, the CMSIS core specification contains some useful debug support. As part of their hardware debug system, the Cortex-M3 and Cortex-M4 provide an instrumentation trace unit (ITM). This can be thought of as a debug UART that is connected to a console window in the debugger. By adding debug hooks (instrumenting) into your code it is possible to read and write data to the debugger while the code is running. We will look at using the instrumentation trace for additional debug and software testing in Chapter 8. For now there are a couple of CMSIS functions that standardize communication with the ITM.

Exercise: Simple ITM

In this exercise we will look at the use of the instrumentation trace in a Cortex-M3 microcontroller. This is an exercise involving hardware debug so you will need the STM32 discovery board or other evaluation board.

Connect the discovery board to the PC via its USB debug port.

Open the exercise in c:\exercises\CMSIS core debug.

In this exercise the code configures the instrumentation trace and then reads and writes characters from the instrumentation port.

Build the project and start the debugger.

Table 4.9: CMSIS Debug Functions

CMSIS Debug Function	Description
volatile int ITM_RxBuffer = ITM_RXBUFFER_EMPTY;	Declare one word of storage for receive flag
ITM_SendChar(c);	Send one character to the ITM
ITM_CheckChar()	Check if any data has been received
ITM_ReceiveChar()	Read one character from the ITM

Open the View\Serial Windows\Debug (printf) Viewer window.

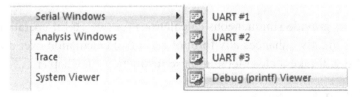

Start running the code.

As the code runs it will send characters to the debug viewer.

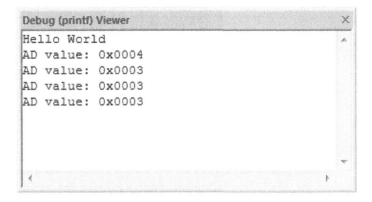

Highlight the debug viewer, enter menu options, and check they are received in the code.

The ITM allows you to use two-way communication with the target software without the need to use any of the microcontroller peripherals. The ITM can be very useful for teasing out more complex debug problems and can also be used for functional software testing. We will look at configuring the ITM and also how to use it for software testing in Chapter 8.

CHAPTER 5

Advanced Architecture Features

Introduction

In the last few chapters, we have covered most of what you need to know to develop a Cortex-M-based microcontroller. In the remaining chapters in this book, we will look at some of the more advanced features of the Cortex-M processors. In this chapter, we will look at the different operating modes built into each of the Cortex-M processors and some additional instructions that are designed to support the use of an RTOS. We will also have a look at the optional MPU, which can be fitted to the Cortex-M3, Cortex-M4, and Cortex-M0+ , and how this can partition the memory map. This provides controlled access to different regions of memory depending on the running processor mode. To round the chapter off, we will have a look at the bus interface between the Cortex-M processor and the microcontroller system.

Cortex Processor Operating Modes

When the Cortex-M processor comes out of reset, it is running in a simple "flat" mode where all of the application code has access to the full processor address space and unrestricted access to the CPU and NVIC registers. While this is OK for many applications, the Cortex-M processor has a number of features that let you place the processor into a more advanced operating mode that is suitable for high-integrity software and also supports an RTOS.

As a first step to understanding the more advanced operating modes of the Cortex-M processor, we need to understand its operating modes. The CPU can be running in two different modes, thread mode and handler mode. When the processor is executing background code (i.e., noninterrupt code), it is running in thread mode. When the processor is executing interrupt code, it is running in handler mode.

		Operations (Privilege out of reset)	Stacks (Main out of reset)
Modes (Thread out of reset)	Handler - Processing of exceptions	Privileged execution full control	Main stack used by OS and exceptions
	Thread - No exception is being processed - Normal code execution	Privileged or unprivileged	Main or process

Figure 5.1
Each Cortex-M processor has two execution modes—(interrupt) handler and (background) thread. It is possible to configure these modes to have privileged and unprivileged access to memory regions. It is also possible to configure a two-stack operating mode.

When the processor starts to run out of reset, there is no operating difference between thread and handler modes. Both modes have full access to all features of the CPU. This is known as privileged mode. By programming the Cortex-M processor CONTROL register, it is possible to place the thread mode in unprivileged mode by setting the thread privilege level bit.

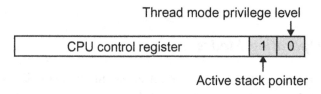

Figure 5.2
The CONTROL register is a CPU register that can only be accessed by the MRS and MSR instructions. It contains two bits that configure the thread mode privilege level and activation of the process.

In unprivileged mode, the MRS, MSR, and Change Processor State (CPS) instructions are disabled for all special CPU registers, except the APSR. This prevents the Cortex-M processor from accessing the CONTROL, FAULTMASK, and PRIMASK registers, and the program status register (except the APSR). In unprivileged mode, it is also not possible to access the systick timer registers, NVIC, or the system control block. These limits attempt to prevent unprivileged code from accidentally disturbing the operation of the Cortex-M processor. If thread mode has been limited to unprivileged access, it is not possible to clear the thread mode privilege level bit even if the CPU is running in handler mode with privilege access. Once the TLP bit has been set, the application code running in thread mode can no longer influence the operation of the Cortex-M processor. When the processor responds to an exception or interrupt, it moves into handler mode, which always executes code in privileged mode regardless of the contents of the CONTROL register. The CONTROL register also contains an additional bit, the active stack pointer selection (ASPEL). Setting this bit enables an additional stack called the process stack. The CONTROL register is a CPU register rather than a memory-mapped register and can only be accessed by the MRS and MSR instructions. The CMSIS core specification provides dedicated functions to read and write to the CONRTOL register.

```
void   __set_CONTROL(uint32_t value);
uint32_t  __get_CONTROL(void);
```

The process stack is a banked R13 stack pointer, which is used by code running in thread mode. When the Cortex-M processor responds to an exception it enters handler mode. This causes the CPU to switch stack pointers. This means that the handler mode will use the MSP while thread mode uses the process stack pointer.

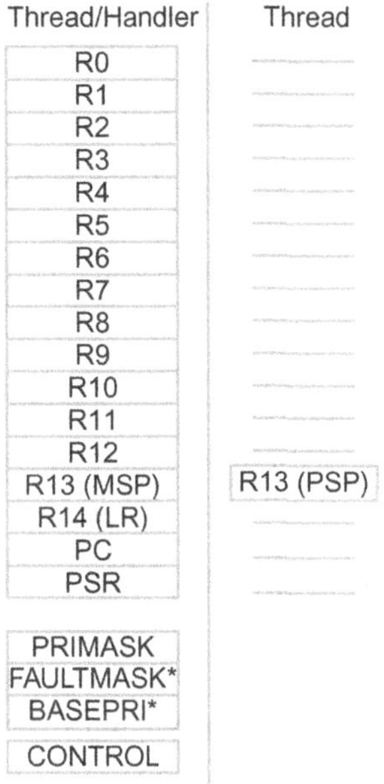

Thread/Handler	Thread
R0	
R1	
R2	
R3	
R4	
R5	
R6	
R7	
R8	
R9	
R10	
R11	
R12	
R13 (MSP)	R13 (PSP)
R14 (LR)	
PC	
PSR	
PRIMASK	
FAULTMASK*	
BASEPRI*	
CONTROL	

Figure 5.3
At reset R13 is the MSP and is automatically loaded with the initial stack value. The CPU CONTROL register can be used to enable a second banked R13 register. This is the process stack that is used in thread mode. The application code must load an initial stack value into this register.

As we have seen in Chapter 3, at reset the MSP will be loaded with the value stored in the first 4 bytes of memory. However, the application stack is not automatically initialized and must be set up by the application code before it is enabled. Fortunately, the CMSIS core specification contains a function to configure the process stack.

```
void   __set_PSP(uint32_t TopOfProcStack);
uint32_t  __get_PSP(void);
```

So if you have to manually set the initial value of the process stack, what should it be? There is not an easy way to answer this, but the compiler produces a report file that details the static calling tree for the project. This file is created each time the project is built and is called <project name>.htm. The report file includes a value for the maximum stack usage and a calling tree for the longest call chain.

```
Stack_Size EQU 0x00000400
  AREA STACK, NOINIT, READWRITE, ALIGN=3
Stack_Mem SPACE Stack_Size
__initial_sp
```

Option	Value
⊟ Stack Configuration	
Stack Size (in Bytes)	0x400
⊟ Heap Configuration	
Heap Size (in Bytes)	0x0

Figure 5.4
The stack size allocated to the MSP is defined in the startup code and can be configured through the configuration wizard.

This calling tree is likely to be for background functions and will be the maximum value for the process stack pointer. This value can also be used as a starting point for the MSP.

Exercise: Stack Configuration

In this exercise, we will have a look at configuring the operating mode of the Cortex-M processor so the thread mode is running with unprivileged access and uses the process stack pointer.

Open the project in the exercise\operating mode folder.

Build the code and start the debugger.

This is a version of the blinky project we used earlier with some code added to configure the processor operating mode. The new code includes a set of #defines.

```
#define USE_PSP_IN_THREAD_MODE      (1<<1)
#define THREAD_MODE_IS_UNPRIVILIGED 1
#define PSP_STACK_SIZE              0x200
```

The first two declarations define the location of the bits that need to be set in the control register to enable the process stack and switch the thread mode into unprivileged access. Then, we define the size of the process stack space. At the start of main, we can use the CMSIS functions to configure and enable the process stack. We can also examine the operating modes of the processor in the register window.

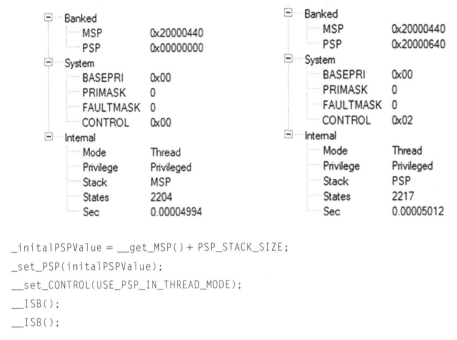

```
_initalPSPValue = __get_MSP() + PSP_STACK_SIZE;
_set_PSP(initalPSPValue);
__set_CONTROL(USE_PSP_IN_THREAD_MODE);
__ISB();
__ISB();
```

When you reach the main() function, the processor is in thread mode with full privileged access to all features of the microcontroller. Also only the MSP is being used. If you step through the three configuration lines, the code first reads the contents of the MSP. This will be at the top of the main stack space. To get the start address for the process stack, we simply add the desired stack size in bytes. This value is written by the process stack before enabling it in the control register. Always configure the stack before enabling it in case there is an active interrupt that could occur before the stack is ready. Next, we need to execute any code that needs to configure the processor before switching the thread mode to unprivileged access. The ADC_Init function accesses the NVIC to configure an interrupt, and the systick timer is also configured. Accessing these registers will be prohibited when we switch to unprivileged mode. Again an instruction barrier is used to ensure the code completes before execution continues.

```
ADC_Init();
  SysTick_Config(SystemCoreClock / 100);
    __set_CONTROL(USE_PSP_IN_THREAD_MODE
            |THREAD_MODE_IS_UNPRIVILIGED);
  __ISB();
```

Banked		
MSP	0x20000440	
PSP	0x20000640	
System		
BASEPRI	0x00	
PRIMASK	0	
FAULTMASK	0	
CONTROL	0x03	
Internal		
Mode	Thread	
Privilege	Unprivileged	
Stack	PSP	
States	202601	
Sec	0.00283324	

Next, set a breakpoint in the IRQ.c module, line 32. This is in the SysTick interrupt handler routine. Now run the code and it will hit the breakpoint when the systick handler interrupt is raised.

```
27 ⊟void SysTick_Handler (void) {
28      static unsigned long ticks = 0;
29      static unsigned long timetick;
30      static unsigned int  leds = 0x01;
31
32 ⊟    if (ticks++ >= 99) {
33        ticks    = 0;
34        clock_1s = 1;
35      }
```

Banked	
MSP	0x20000438
PSP	0x20000620
System	
BASEPRI	0x00
PRIMASK	0
FAULTMASK	0
CONTROL	0x01
Internal	
Mode	Handler
Privilege	Privileged
Stack	MSP
States	1642612
Sec	0.02283339

Now that the processor is serving an interrupt, it has moved into interrupt handler mode with privileged access to the Cortex processor and is using the main stack.

Supervisor Call

Once configured, this more advanced operating mode provides a partition between the exception/interrupt code running in handler mode and the background application code running in thread mode. Each operating mode can have its own code region, RAM region, and stack. This allows the interrupt handler code full access to the chip without the risk that it may be corrupted by the application code. However, at some point, the application code will need to access features of the Cortex-M processor that are only available in the handler mode with its full privileged access. To allow this to happen, the Thumb-2 instruction set has an instruction called Supervisor Call (SVC). When this instruction is executed, it raises a supervisor exception that moves the processor from executing the application code in thread\unprivileged mode to an exception routine in handler/privileged mode. The SVC has its own location within the vector table and behaves like any other exception.

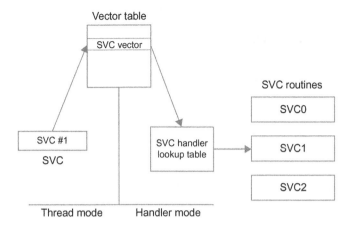

Figure 5.5
The SVC allows execution to move from unprivileged thread mode to privileged handler mode and gain full unrestricted access to the Cortex processor. The SVC instruction is used by RTOS API calls.

The supervisor instruction may also be encoded with an 8-bit value called an ordinal. When the SVC is executed, this ordinal value can be read and used as an index to call one of 256 different supervisor functions.

Figure 5.6
The unused portion of the SVC instruction can be encoded with an ordinal number. On entry to the SVC handler, this number can be read to determine which SVC functions to execute.

The toolchain provides a dedicated SVC support function that is used to extract the ordinal value and call the appropriate function. First, we need to find the value of the PC, which is the last value stored on the stack. First, the SVC support function reads the link register to determine the operating mode; then it reads the value of the saved PC from the appropriate stack. We can then read the memory location holding the SVC instruction and extract the ordinal value. This number is then used as an index into a lookup table to load the address of the function that is being called. We can then call the supervisor function that is executed in privileged mode before we return back to the application code running in unprivileged thread mode. This mechanism may seem an overly complicated way of calling a function, but it provides the basis of a supervisor/user split where an OS is running in privileged mode and acts as a supervisor to application threads running in unprivileged thread mode. This way the individual threads do not have access to critical processor features except by making API calls to the OS.

Exercise: SVC

In this exercise, we will look at calling some functions with the SVC instruction rather than branching to the routine as in a standard function call. Open the project in the exercise/SVC folder. First, let us have a look at the project structure.

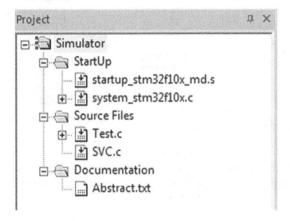

The project consists of the standard project startup file and the initializing system file. The application source code is in the file Test.c. There is an additional source file SVC.c that provides support for handling SVC exceptions. The SVC.c file contains the SVC exception handler; this is a standard support file that is provided with the ARM compiler. We will have a closer look at its operation later. The application code in Test.c is calling two simple functions that in turn call routines to perform basic arithmetic operations.

```
int main (void) {
    test_a();
    test_t();
while(1);
}
  void test_a (void) {
    res = add (74, 27);
    res + = mul4(res);
}
void test_t (void) {
    res = div (res, 10);
    res = mod (res, 3);
}
```

Each of the arithmetic functions is designed to be called with an SVC instruction so that all of these functions run in handler mode rather than thread mode. In order to convert the arithmetic functions from standard functions to software interrupt functions, we need to change the way the function prototype is declared. The way this is done will vary between compilers, but in the ARM compiler there is a function qualifier __svc. This is used as shown below to convert the function to an SVC and allows you to pass up to four parameters and get a return value. So the add function is declared as follows:

```
int __svc(0) add (int i1, int i2);
int __SVC_0  (int i1, int i2) {
  return (i1 + i2);
}
```

The __svc qualifier defines this function as an SVC and defines the ordinal number of the function. The ordinals used must start from zero and grow upward contiguously to a maximum of 256. To enable each ordinal, it is necessary to build a lookup table in the SVC.c file.

```
; Import user SVC functions here.
  IMPORT   __SVC_0
  IMPORT   __SVC_1
  IMPORT   __SVC_2
  IMPORT   __SVC_3
```

SVC_Table

```
; Insert user SVC functions here
  DCD   __SVC_0
  DCD   __SVC_1  ;
  DCD   __SVC_2  ;
  DCD   __SVC_3  ;
```

You must import the label used for each supervisor function and then add the function labels to the SVC table. When the code is compiled, the labels will be replaced by the entry address of each function.

In the project build the code and start the simulator. Step the code until you reach line 61, the call to the add function.

The following code is displayed in the disassembly window, and in the register window we can see that the processor is running in thread mode.

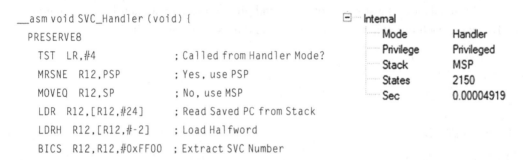

```
61: res = add (74, 27);
0x0800038A 211B  MOVS  r1,#0x1B
0x0800038C 204A  MOVS  r0,#0x4A
0x0800038E DF03  SVC   0x03
```

```
Internal
  Mode        Thread
  Privilege   Privileged
  Stack       MSP
  States      2120
  Sec         0.00004878
```

The function parameters are loaded into the parameter passing registers R0 and R1, and the normal branch instruction is replaced by an SVC instruction. The SVC instruction is encoded with an ordinal value of 3. If you make the disassembly window the active window and step through these instructions, the SVC exception will be raised and you will enter the SVC_Handler in SVC.c. In the registers window, you can also see that the processor is now running in handler mode.

```
__asm void SVC_Handler (void) {
  PRESERVE8
    TST  LR,#4          ; Called from Handler Mode?
    MRSNE R12,PSP       ; Yes, use PSP
    MOVEQ R12,SP        ; No, use MSP
    LDR  R12,[R12,#24]  ; Read Saved PC from Stack
    LDRH R12,[R12,#-2]  ; Load Halfword
    BICS R12,R12,#0xFF00 ; Extract SVC Number
```

```
Internal
  Mode        Handler
  Privilege   Privileged
  Stack       MSP
  States      2150
  Sec         0.00004919
```

The first section of the SVC_Handler code works out which stack is in use and then reads the value of the program counter saved on the stack. The program counter value is the return address, so the load instruction deducts 2 to get the address of the SVC instruction. This will be the address of the SVC instruction that raised the exception. The SVC instruction is then loaded into R12 and the ordinal value is extracted. The code is using R12 because the ARM binary interface standard defines R12 as the "intra Procedure call scratch register"; this means it will not contain any program data and is free for use.

```
  PUSH {R4,LR}               ; Save Registers
    LDR  LR, = SVC_Count
    LDR  LR,[LR]
    CMP  R12,LR
    BHS  SVC_Dead            ; Overflow
    LDR  LR, = SVC_Table
    LDR  R12,[LR,R12,LSL #2] ; Load SVC Function Address
    BLX  R12                 ; Call SVC Function
```

The next section of the SVC exception handler prepares to jump to the add() function. First, the link register and R4 are pushed onto the stack. The size of the SVC table is loaded into the link register. The SVC ordinal is compared to the table size to check whether it is less than the SVC table size and hence a valid number. If it is valid, the function address is loaded into R12 from the SVC table and the function is called. If the ordinal number has not been added to the table, the code will jump to a trap called SVC_DEAD. Although R4 is not used in this example, it is preserved on the stack as it is possible for the called function to use it.

```
POP  {R4,LR}
TST  LR,#4
MRSNE R12,PSP
MOVEQ R12,SP
STM  R12,{R0-R3}  ; Function return values
BX  LR  ; RETI
```

Once the SVC function has been executed, it will return back to the SVC exception handler to clean up before returning to the background code in handler mode.

Pend_SVC Exception

To extend the support for OSes, there is a second software interrupt called Pend_SVC. The purpose of the Pend_SVC exception is to minimize the latency experienced by interrupt service routines running alongside the OS. A typical OS running on the Cortex-M processor will use the SVC exception to allow user threads to make calls to the OS API, and the systick timer will be used to generate a periodic exception to run the OS scheduler.

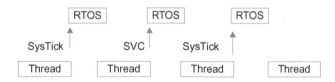

Figure 5.7
A typical RTOS will use the systick timer to provide a periodic interrupt to run
the kernel scheduler. Calls to RTOS functions will use the SVC instruction to
raise an exception.

While this works fine, problems can occur when the systick timer is running at a high priority. If we enter a peripheral interrupt and start to run the ISR and then the systick timer interrupt is triggered, it will preempt the peripheral interrupt and delay servicing of the peripheral interrupt while the OS scheduler examines the state of the user threads and performs a context switch to start a new thread running. This process can delay execution of the peripheral interrupt by several hundred cycles.

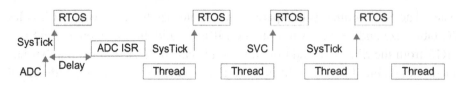

Figure 5.8

If the RTOS systick interrupt is running at a high priority, it can delay peripheral interrupts making interrupt handling nondeterministic.

The Pend_SVC is used to minimize this problem. When the systick timer occurs, it will check the NVIC registers to see if any peripheral interrupt is pending. If this is the case, it will cause a Pend_SVC exception by setting the Pend bit in the NVIC Pend_SVC register. The OS scheduler then exits. Now we have two active interrupts, the original peripheral interrupt and the Pend_SVC exception. The Pend_SVC is set to a low priority so that the peripheral interrupt ISR will resume execution. As soon as the peripheral interrupt has finished, the Pend_SVC exception will be served, which will continue with the OS scheduler to finish the thread context switch. This technique maintains the performance of the OS while keeping the intrusion on the peripheral interrupts to a minimum.

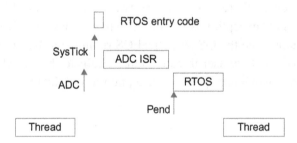

Figure 5.9

When execution enters the RTOS scheduler, it can check if any other interrupts are pending. If this is the case, the low-priority Pend_SVC can be set to pend and the RTOS scheduler suspended. The peripheral interrupt will now be served. When it is finished the Pend_SVC exception will be served, allowing the RTOS scheduler to resume processing.

Example: Pend_SVC

In this example, we will examine how to use the Pend_SVC system service call interrupt.

Open the project in c:\examples\Pendsvc.

Build the project and start the debugger.

The code initializes the ADC and enables its end of conversion interrupt. It also changes the Pend_SVC and ADC interrupt priority from their default options. The SVC has priority zero (highest) while the ADC has priority 1, and the Pend_SVC interrupt has priority 2 (lowest). The systemCode routine uses an SVC instruction to raise an exception and move to handler mode.

```
NVIC_SetPriority(PendSV_IRQn,2);  //set interrupt priorities
NVIC_SetPriority(ADC1_2_IRQn,1);
NVIC_EnableIRQ(ADC1_2_IRQn);      //enable the ADC interrupt
ADC1->CR1  |= (1UL << 5);         //switch on the ADC and start a conversion
ADC1->CR2  |= (1UL << 0);
ADC1->CR2  |= (1UL << 22);
systemCode();                     //call some system code with an SVC interrupt
```

Set a breakpoint on the systemCode( ) function and run the code.

```
58 | ADC1->CR2    |=  (1UL << 22);
59 | systemCode();
60 |
```

Open the Peripherals\Core Peripherals\Nested Vector Interrupt Controller window and check the priority levels of the SVC, Pend_SVC, and ADC interrupts.

Idx	Source	Name	E	P	A	Priority
11	System Service Call	SVCALL	1	0	0	0
14	Pend System Service	PENDSV	1	0	0	2
34	ADC Global Interrupt	ADC	1	0	0	1

Now step into the systemCode routine (F11) until you reach the C function.

```
void __svc(0) systemCode (void);
void __SVC_0 (void) {
unsigned int i, pending;
  for(i = 0;i < 100;i++);
pending = NVIC_GetPendingIRQ(ADC1_2_IRQn);
if(pending == 1){
   SCB->ICSR |= 1<<28;           //set the pend pend
}else{
  Do_System_Code();
}
}
```

Inside the system code routine, there is a short loop that represents the critical section of code that must be run. While this loop is running, the ADC will finish conversion, and as it has a lower priority than the SVC interrupt it will enter a pending state. When we exit the loop, we test the state of any critical interrupts by reading their pending bits. If a critical interrupt is pending, then the remainder of the system code routine can be delayed. To do this, we set the PENDSVSET bit in the interrupt control and state register and quit the SVC handler.

Set a breakpoint on the exit brace (}) of the systemCode routine and run the code.

```
38        Do_System_Code();
39    } }
40    }
```

Now use the NVIC debug window to examine the state of the interrupts.

Idx	Source	Name	E	P	A	Priority
11	System Service Call	SVCALL	1	0	1	0
14	Pend System Service	PENDSV	1	1	0	2
34	ADC Global Interrupt	ADC	1	1	0	1

Now the SVC call is active with the ADC and Pend_SVC system service call in a pending state.

Single Step (F11) to the end of the system service call. Continue to single step so that you exit the system service routine and enter the next pending interrupt.

Both of the pending interrupts will be tail chained on to the end of the system service call. The ADC has the highest priority so it will be served next.

Idx	Source	Name	E	P	A	Priority
11	System Service Call	SVCALL	1	0	0	0
14	Pend System Service	PENDSV	1	1	0	2
34	ADC Global Interrupt	ADC	1	0	1	1

Step out of the ADC handler and you will immediately enter the Pend_SVC system service interrupt, which allows you to resume execution of the system code that was requested to be executed in the system service call interrupt.

Interprocessor Events

The Cortex-M processors are designed so that it is possible to build multiprocessor devices. An example would be to have a Cortex-M4 and a Cortex-M0 within the same microcontroller. The Cortex-M0 will typically manage the user peripherals, while the Cortex-M4 runs the intensive portions of the application code. Alternatively, there are devices that have a Cortex-A8 that can run Linux and manage a complex user interface;

on the same chip there are two Cortex-M4s that manage the real-time code. These more complex system-on-chip designs require methods of signaling activity between the different processors. The Cortex-M processors can be chained together by an external event signal. The event signal is set by using a set event instruction. This instruction can be added to your C code using the __SEV() intrinsic provided by the CMSIS core specification. When a SEV instruction is issued it will wake up the target processor if it has entered a low-power mode using the WFE instruction. If the target processor is running, the event latch will be set so that when the target processor executes the WFE instruction it will reset low-power mode.

Exclusive Access

One of the key features of an RTOS is multitasking support. As we will see in the next chapter, this allows you to develop your code as independent threads that conceptually are running in parallel on the Cortex-M processor. As your code develops, the program threads will often need to access common resources, be it SRAM or peripherals. An RTOS provides mechanisms called semaphores and mutexes that are used to control access to peripherals and common memory objects.

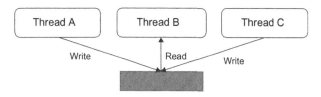

Figure 5.10

In a multiprocessor or multithread environment, it is necessary to control access to shared resources, or errors such as read before write can occur.

While it is possible to design "memory lock" routines on any processor, the Cortex-M3 and Cortex-M4 provide a set of instructions that can be used to optimize exclusive access routines.

Table 5.1: Exclusive Access Instructions

__LDREXB	Load exclusive (8 bits)
__LDREXH	Load exclusive (16 bits)
__LDREXW	Load exclusive (32 bits)
__STREXB	Store exclusive (8 bits)
__STREXH	Store exclusive (16 bits)
__STREXW	Store exclusive (32 bits)
__CLREX	Remove exclusive lock

In earlier ARM processors like the ARM7 and ARM9, the problem of exclusive access was answered by a swap instruction that could be used to exchange the contents of two registers. This instruction took four cycles but it was an atomic instruction, meaning that once started it could not be interrupted and was guaranteed exclusive access to the CPU to carry out its operation. As Cortex-M processors have multiple busses, it is possible for read and write accesses to be carried out on different busses and even by different bus masters, which may themselves be additional Cortex-M processors. On the Cortex-M processor, the new technique of exclusive access instructions has been introduced to support multitasking and multiprocessor environments.

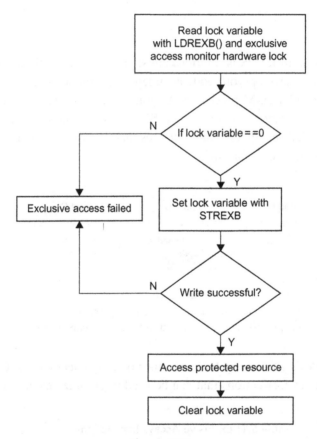

Figure 5.11
The load and store exclusive instructions can be used to control access to a memory resource.
They are designed to work with single and multiprocessor devices.

The exclusive access system works by defining a lock variable to protect the shared resource. Before the shared resource can be accessed, the locked variable is checked using the exclusive read instruction; if it is zero, then the shared resource is not currently being accessed. Before we access the shared resource, the lock variable must be set using the exclusive store instruction. Once the lock variable has been set, we now have control of the shared resource and can write to it. If our process is preempted by an interrupt or another thread that also performs an exclusive access read, then a hardware lock in the exclusive access monitor is set, preventing the original exclusive store instruction from writing to the lock variable. This gives exclusive control to the preempting process. When we are finished with the shared resource, the lock variable must be written to zero; this clears the variable and also removes the lock. If your code starts the exclusive access process but needs to abandon it, there is a clear exclusive (CLREX) instruction that can be used to remove the lock. The exclusive access instructions control access between different processes running on a single Cortex-M processor, but the same technique can be extended to a multiprocessor environment provided that the silicon designer includes the additional monitor hardware bus signals.

Exercise: Exclusive Access

In this exercise, we will create an exclusive access lock that is shared between a background thread process and an SVC handler routine to demonstrate the lock and unlock process.

Open the exercise in c:\exercises\exclusive access.

Build the code and start the debugger.

```
int main (void) {
if(__LDREXB( &lock_bit) = = 0){
  if (!__STREXB(1,&lock_bit = =0)){
    semaphore++;
    lock_bit = 0;
      }
    }
```

The first block of code demonstrates a simple use of the exclusive access instructions. We first test the lock variable with the exclusive load instruction. If the resource is not locked by another process, we set the lock bit with the exclusive store instruction. If this is successful, we can then access the shared memory resource called a semaphore. Once this variable has been updated, the lock variable is written to zero, and the clear exclusive instruction releases the hardware lock.

Step through the code to observe its behavior.

```
if(__LDREXB( &lock_bit)==0){
    thread_lock();
  if (!__STREXB(1,&lock_bit)){
    semaphore++;
  }
```

The second block of code does exactly the same thing except between the exclusive load and exclusive store the function is called thread_lock(). This is an SVC routine that will enter handler mode and jump to the SVC0 routine.

```
void __svc(0) thread_lock (void);
void __SVC_0  (void) {
__LDREXB( &lock_bit);
}
```

The SVC routine simply does another exclusive read of the lock variable that will set the hardware lock in the exclusive access monitor. When we return to the original routine and try to execute the exclusive store instruction, it will fail because any exception that happens between LDREX and STREX will cause the STREX to fail. The local exclusive access monitor is cleared automatically at exception entry/exit.

Step through the second block of code and observe the lock process.

Memory Protection Unit

The Cortex-M3/4 and Cortex-M0+ processors have an optional memory protection hardware unit that may be included in the processor core by the silicon manufacturer when the microcontroller is designed. The MPU allows you to extend the privileged/ unprivileged code model. If it is fitted, the MPU allows you to define regions within the memory map of the Cortex-M processor and grant privileged or unprivileged access to these regions. If the processor is running in unprivileged mode and tries to access an address within a privileged region, a memory protection exception will be raised and the processor will vector to the memory protection ISR. This allows you to detect and correct runtime memory errors.

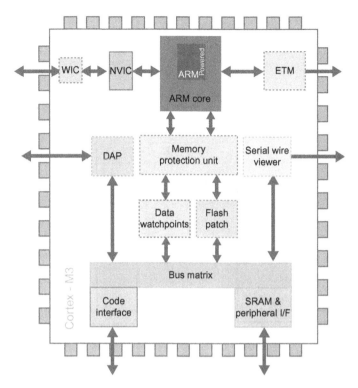

Figure 5.12
The MPU is available on the Cortex-M0+, Cortex-M3, and Cortex-M4. It allows you to place a protection template over the processor memory map.

In practice, the MPU allows you to define eight memory regions within the Cortex-M processor address space and grant privileged or unprivileged access to each region. These regions can then be further subdivided into eight equally sized subregions that can in turn be granted privileged or unprivileged access. There is also a default background region that covers the entire 4 GB address space. When this region is enabled, it makes access to all memory locations privileged. To further complicate things, memory regions may be overlapped with the highest numbered region taking precedent. Also any area of memory that is not covered by an MPU region may not have any kind of memory access.

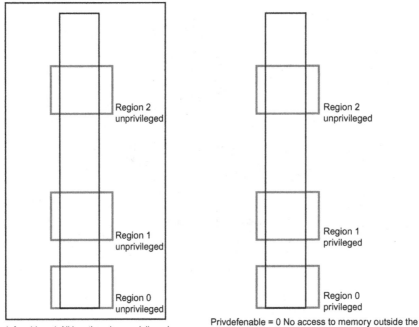

Region 2
unprivileged

Region 1
unprivileged

Region 0
unprivileged

Privdefenable = 1 All locations have privileged access

Region 2
unprivileged

Region 1
privileged

Region 0
privileged

Privdefenable = 0 No access to memory outside the
MPU regions

Figure 5.13
The MPU allows you to define eight regions, each with eight subregions over the processor
memory map. Each region can grant different access privileges to its address range. It is also
possible to set a default privileged access over the whole memory map and then create "holes"
with different access privileges.

So it is possible to build up complex protection templates over the memory address space.
This allows you to design a protection regime that helps build a robust operating
environment but also gives you enough rope to hang yourself.

Configuring the MPU

The MPU is configured through a group of memory-mapped registers located in the Cortex-
M processor system block. These registers may only be accessed when the Cortex processor
is operating in privileged mode.

MPU registers
Control
Region number
Base address
Attribute and size

Figure 5.14
Each MPU region is configured through the region, base address, and attribute registers. Once each region is configured, the control register makes the MPU regions active.

The control register contains three active bits that effect the overall operation of the MPU. The first bit is the PRIVDEFENABLE bit that enables privileged access over the whole 4 GB memory map. The next bit is the HFNMIENA; when set, this bit enables the operation of the MPU during a hard fault, NMI, or FAULTMASK exception. The final bit is the MPU enable bit; when set, this enables the operation of the MPU. Typically, when configuring the MPU, the last operation performed is to set this bit. After reset, all of these bits are cleared to zero.

Figure 5.15
The control register allows you to enable the global privileged region with the Privilege default Enable (PRIVDEFENA). The enable bit is used to make the configured MPU regions active, and Hard Fault and Non Maskable Interrupt Enable (HFNMIEA) bit is used to enable the regions when the hard fault, NMI, or FAULTMASK exceptions are active.

The remaining registers are used to configure the eight MPU regions. To configure a given region (0–7), first select the region by writing its region number into the region number register. Once a region has been selected, it can then be configured by the base address and the attribute and size register. The base address register contains an address field which consists of the upper 27 address bits, the lower 5 address bits are not supported and are replaced by a valis bit and a repeat of the four bit MPU region number.

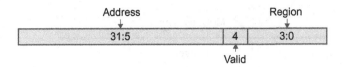

Figure 5.16
The address region of the base address register allows you to set the start address of an MPU region. The address values that can be written will depend on the size setting in the attribute and size register. If valid is set to 1, then the region number set in the region field is used; otherwise the region number in the region register is used.

As you might expect, the base address of the MPU region must be programmed into the address field. However, the available base addresses that may be used depend on the size of the defined for the region. The minimum size for a region is from 32 bytes upto 4 GB. The base address of an MPU region must be a multiple of the region size. Programming the address field sets the selected region's base address. You do not need to set the valid bit. If you write a new region number into the base address register region field, set the valid bit and write a new address; you can start to configure a new region without the need to update the region number register. Programming the attribute and size register finishes configuration of an MPU region.

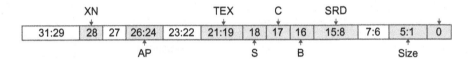

Figure 5.17
The attribute and size register allows you to define the MPU region size from 32 bytes to 4 GB. It also configures the memory attributes and access control options.

The size field defines the address size of the memory protection region in bytes. The region size is calculated using the formula

$$\text{MPU region memory size} = 2 \, \text{POW}(\text{SIZE} + 1)$$

This gives us a minimum size starting at just 32 bytes. As noted above, the selected size also defines the range of possible base addresses. Next, it is possible to set the region attributes and access privileges. Like the Cortex-M processor, the MPU is designed to support multiprocessor systems. Consequently, it is possible to define regions as being shared between Cortex-M processors or as being exclusive to the given processor.

It is also possible to define the cache policy for the area of memory covered by the MPU region. Currently, the vast majority of microcontrollers only have a single Cortex processor, although asymmetrical multiprocessor devices have started to appear (Cortex-M4 and Cortex-M0), and currently no microcontroller has a cache. The attributes are defined by the TEX, C, B, and S bits, and suitable settings for most microcontrollers are shown below.

Table 5.2: Memory Region Attributes

Memory Region	TEX	C	B	S	Attributes
Flash	000	1	0	0	Normal memory, nonshareable, write through
Internal SRAM	000	1	0	1	Normal memory, shareable, write through
External SRAM	000	1	1	1	Normal memory, shareable, write back, write allocate
Peripherals	000	0	1	1	Device memory, shareable

When working with the MPU, we are more interested in defining the access permissions. These are defined for each region in the AP field.

Table 5.3: Memory Access Rights

AP	Privileged	Unprivileged	Description
000	No access	No access	All accesses generate a permission fault
001	RW	No access	Access from privileged code only
010	RW	RO	Unprivileged writes cause a permission fault
011	RW	RW	Full access
100	Unpredictable	Unpredictable	Reserved
101	RO	No access	
110	RO	RO	Reads by privileged code only
111	RO	RO	Read only for privileged and unprivileged code

Once the size, attributes, and access permissions are defined, the enable bit can be set to make the region active. When each of the required regions has been defined, the MPU can be activated by setting the global enable bit in the control register. When the MPU is active and the application code makes an access that violates the permissions of a region, an MPU exception will be raised. Once you enter an MPU exception, there are a couple of registers that provide information to help diagnose the problem. The first byte of the configurable fault status register located in the system control block is called the memory manager fault status register.

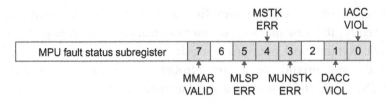

Figure 5.18

The MPU fault status register is a subsection of the configurable fault status register in the system control block. It contains error flags that are set when an MPU exception occurs. The purpose of each flag is shown.

Table 5.4: Fault Status Register Flag Descriptions

Flag	Description
IACCVIOL	Instruction access violation status flag
DACCVIOL	Data access violation status flag
MUNSTKERR	Memory manager fault on unstacking
MSTKERR	Memory manager fault on stacking
MLSPERR	Memory manager FPU lazy stacking error, Cortex-M4 only (see Chapter 7)
MMARVALID	Memory manager fault address valid

Depending on the status of the fault conditions, the address of the instruction that caused the memory fault may be written to a second register, that is, the memory manager fault address register. If this address is valid, it may be used to help diagnose the fault.

Exercise: MPU Configuration

In this exercise, we will configure the MPU to work with the blinky project.

Open the project in exercises\Blinky MPU.

This time, the microcontroller used is an NXP LPC1768, which has a Cortex-M3 processor fitted with the MPU. First, we have to configure the project so that there are distinct regions of code and data that will be used by the processor in thread and handler modes. When doing this, it is useful to sketch out the memory map of the application code and then define a matching MPU template.

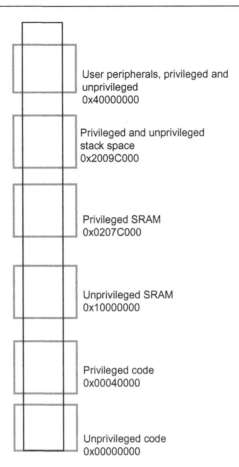

User peripherals, privileged and
unprivileged
0x40000000

Privileged and unprivileged
stack space
0x2009C000

Privileged SRAM
0x0207C000

Unprivileged SRAM
0x10000000

Privileged code
0x00040000

Unprivileged code
0x00000000

Figure 5.19
The memory map of the blinky project can be split into six regions.

Region 0: Unprivileged application code
Region 1: Privileged system code
Region 2: Unprivileged SRAM
Region 3: Privileged SRAM
Region 4: Privileged and unprivileged stack space
Region 5: Privileged and unprivileged user peripherals

Now open the Options for Target\Target menu. Here, we can set up the memory regions to match the proposed MPU protection template.

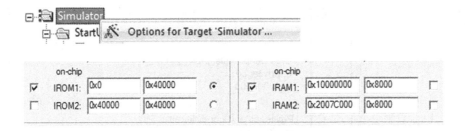

The target memory map defines two regions of code memory: 0-0x3FFF and 0x4000–0x7FFF. The lower region will be used as the default region to hold the application code and will be accessed by the processor in unprivileged mode. The upper region will be used to hold the interrupt and exception service routines and will be accessed by the processor in privileged mode. Similarly, there are two RAM regions—0x10000000 and 0x100008000—that will hold data used by the unprivileged code and the system stacks, while the upper region—0x0207C000—will be used to hold the data used by the privileged code. In this example, we are not going to set the background privileged region, so we must map MPU regions for the peripherals. All the peripherals except the GPIO are in one contiguous block from 0x40000000, while the GPIO registers sit at 0x000002C9. The peripherals will be accessed by the processor while it is in both privileged and unprivileged modes.

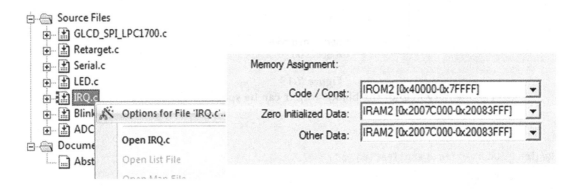

To prepare the code, we need to force the interrupt handler code into the regions that will be granted privileged access. In this example, all the code that will run in handler mode has been placed in one module. In the local options for this module, we can select the code and data regions that will be given privileged access rights by the MPU. All of the other code and data will be placed in the default memory regions that will run in unprivileged mode.

Once the project memory layout has been defined, we can add code to the project to set up the MPU protection template.

```
#define SIZE_FIELD                    1
#define ATTRIBUTE_FIELD               16
#define ACCESS_FIELD                  24
#define ENABLE                        1
#define ATTRIBUTE_FLASH               0x4
#define ATTRIBUTE_SRAM                0x5
#define ATTRIBUTE_PERIPHERAL          0x3
#define PRIV_RW_UPRIV_RW              3
#define PRIV_RO_UPRIV_NONE            5
#define PRIV_RO_UPRIV_RO              6
#define PRIV_RW_UPRIV_RO              2
#define USE_PSP_IN_THREAD_MODE        2
#define THREAD_MODE_IS_UNPRIVILIGED   1
#define PSP_STACK_SIZE                0x200
#define TOP_OF_THREAD_RAM             0x10007FF0
MPU->RNR    =  0x00000000;
MPU->RBAR   =  0x00000000;
MPU->RASR   =  (PRIV_RO_UPRIV_RO<<ACCESS_FIELD)
               |(ATTRIBUTE_FLASH<<ATTRIBUTE_FIELD)
               |(17<<SIZE_FIELD)|ENABLE;
```

The code shown above is used to set the MPU region for the unprivileged thread code at the start of memory. First, we need to set a region number followed by the base address of the region. Since this will be flash memory, we can use the standard attribute for this memory type. Next we can define its access type. In this case, we can grant read-only access for both privileged and unprivileged modes. Next, we can set the size of the region, which is 256 K, must equal 2POW(SIZE + 1), and equate to 17. The enable bit is set to activate this region when the MPU is fully enabled. Each of the other regions are programmed in a similar fashion. Finally, the memory management exception and the MPU are enabled.

```
NVIC_EnableIRQ(MemoryManagement_IRQn);
MPU->CTRL = ENABLE;
```

Start the debugger, set a breakpoint on line 82, and run the code. When the breakpoint is reached, we can view the MPU configuration via the Peripherals\Core Peripherals\Memory Protection Unit menu.

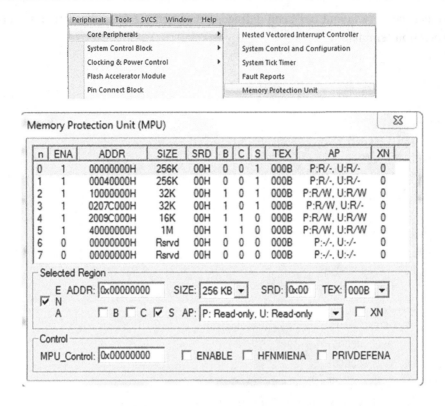

Here, we can easily see the regions defined and the access rights that have been granted.

Now run the code for a few seconds and then halt the processor.

An MPU exception has been raised and execution has jumped to the memManager handler.

```
147   MemManage_Handler\
148                PROC
149                EXPORT   MemManage_Handler          [WEAK]
150                B        .
151                ENDP
```

The question now is what caused the MPU exception? We can find this out by looking at the memory manager fault and status register. This can be directly viewed in the debugger by opening the Peripherals\Core Peripherals\Fault Reports window.

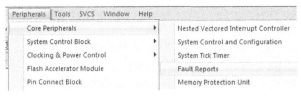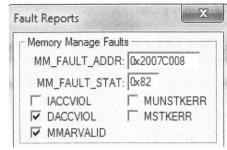

Here, we can see that the fault was caused by a data access violation to address 0x2007C008. If we now open the map file produced by the linker, we can search for this address and find what variable is placed at this location.

Highlight the project name in the project window and double click. This will open the map file. Now use the Edit\Find dialog to search the map file for the address 0x2007C008.

```
579    text            0x10000010   Data    10   blinky.o(.bss)
580    __initial_sp    0x10000220   Data     0   startup_lpc17xx.o(STACK)
581    clock_1s        0x2007c008   Data     1   irq.o(.data)
```

This shows that the variable clock_1s is at 0x2007C008 and that it is declared in irq.c.

Clock_1s is a global variable that is also accessed from the main loop running in unprivileged mode. However, this variable is located in the privileged RAM region, so accessing it while the processor is running in unprivileged mode will cause an MPU fault.

Find the declaration of clock_1s in blinky.c and remove the extern keyword.

Now find the declaration for clock_1s in irq.c and add the keyword extern.

Build the code and view the updated map file.

```
__stdin        0x10000008   Data   4   retarget.o(.data)
clock_1s       0x1000000c   Data   1   blinky.o(.data)
AD_done        0x1000000e   Data   1   adc.o(.data)
```

Now clock_1s is declared in blinky.c and is located in the unprivileged RAM region so that it can be accessed by both privileged and unprivileged code.

Now restart the debugger and the code will run without raising any MPU exceptions.

MPU Subregions

As we have seen, the MPU has a maximum of eight regions that can be individually configured with location size and access type. Any region that is configured with a size of 256 bytes or more will contain eight equally spaced subregions. When the region is configured, each of the subregions is enabled and has the default region attributes and access settings. It is possible to disable a subregion by setting a matching subregion bit in the Subregion Disable (SRD) field of the MPU attribute and size register. When a subregion is disabled, a "hole" is created in the region; this "hole" inherits the attributes and access permission of any overlapped region. If there is no overlapped region, then the global privileged background region will be used. If the background region is not enabled, then no access rights will be granted and an MPU exception will be raised if an access is made to an address in the subregion "hole."

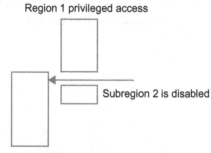

Figure 5.20
Each region has eight subregions. If a subregion is disabled, it inherits the access rights from an overlapped region or the global background region.

If we have two overlapped regions, the region with the highest region number will take precedence. In the case above, an unprivileged region is overlapped by a privileged region. The overlapped section will have privileged access. If a subregion is disabled in region 1, then the access rights in region 0 will be inherited and granted unprivileged access to the subregion range of addresses.

MPU Limitations

When designing your application software to use the MPU, it is necessary to realize that the MPU only monitors the activity of the Cortex-M processor. Many, if not most, Cortex-M-based microcontrollers have other peripherals, such as direct memory access (DMA) units, that are capable of autonomously accessing memory and peripheral registers. These units are additional "bus masters" that arbitrate with the Cortex-M processor to gain access to the microcontroller resources. If such a unit makes an access to a prohibited region of memory, it will not trigger an MPU exception. This is important to remember as the Cortex-M processor has a bus structure that is designed to support multiple independent "bus master" devices.

AHB Lite Bus Interface

The Cortex-M processor family has a final important architectural improvement over the earlier generation of ARM7- and ARM9-based microcontrollers. In these first-generation ARM-based microcontrollers, the CPU was interfaced to the microcontroller through two types of busses. These were the AHB and the advanced peripheral bus (APB).

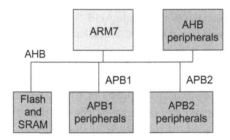

Figure 5.21
The first generation of ARM-based microcontrollers had an internal bus system based on the AHB and the APB. As multiple bus masters (CPU, DMA) were introduced, a bus arbitration phase had to be completed before a transfer could be made across the bus.

The high-speed bus connected the CPU to the flash and SRAM memory while the microcontroller peripherals were connected to one or more APB busses. The AHB bus also supported additional bus masters such as DMA units to sit alongside the ARM7 processor. While this system worked, the bus structure started to become a bottleneck, particularly as more complex peripherals such as Ethernet MAC and USB were added. These peripherals contained their own DMA units, which also needed to act as a bus master. This meant that there could be several devices (ARM7 CPU, general-purpose DMA, and Ethernet MAC) arbitrating for the AHB bus at any given point in time. As more and more complex peripherals are added, the overall throughput and deterministic performance became difficult to predict.

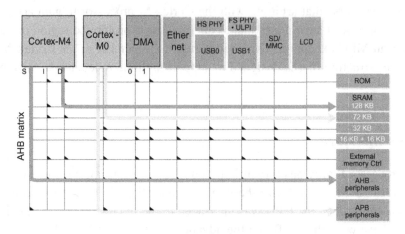

Figure 5.22
The Cortex-M processor family replaces the single AHB bus with a bus matrix that provides parallel paths for each bus master to block each of the slave devices.

The Cortex-M processor family overcomes this problem by using an AHB bus matrix. The AHB bus matrix consists of a number of parallel AHB busses that are connected to different regions of the chip. These regions are laid out by the manufacturer when the chip is designed. Each region is a slave device; this can be the flash memory, a block of SRAM, or a group of user peripherals on an APB bus. Each of these regions is then connected back to each of the bus masters through additional AHB busses to form the bus matrix. This allows manufacturers to design complex devices with multiple Cortex-M processors, DMA units, and advanced peripherals, each with parallel paths to the different device resources. The bus matrix is hardwired into the microcontroller and does not need any configuration by your application code. However, when you are designing the application code, you should pay attention to where different memory objects are located. For example, the memory used by the Ethernet controller should be placed in one block of SRAM while the USB memory is located in a separate SRAM block. This allows the Ethernet, USB and DMA units to work in parallel while the Cortex processor is accessing the flash and user peripherals. So by structuring the memory map of your application code, you can exploit this degree of parallelism and gain an extra boost in performance.

CHAPTER 6

Developing with CMSIS RTOS

Introduction

This chapter is an introduction to using a small footprint RTOS on a Cortex-M microcontroller. If you are used to writing procedural-based C code on small 8/16-bit microcontrollers, you may be doubtful about the need for such an OS. If you are not familiar with using an RTOS in real-time embedded systems, you should read this chapter before dismissing the idea. The use of an RTOS represents a more sophisticated design approach, inherently fostering structured code development that is enforced by the RTOS API.

The RTOS structure allows you to take a more object-orientated design approach, while still programming in C. The RTOS also provides you with multithreaded support on a small microcontroller. These two features actually create quite a shift in design philosophy, moving us away from thinking about procedural C code and flowcharts. Instead we consider the fundamental program threads and the flow of data between them. The use of an RTOS also has several additional benefits that may not be immediately obvious. Since an RTOS-based project is composed of well-defined threads it helps to improve project management, code reuse, and software testing.

The trade-off for this is that an RTOS has additional memory requirements and increased interrupt latency. Typically, the Keil RTX RTOS will require 500 bytes of RAM and 5 KB of code, but remember that some of the RTOS code would be replicated in your program anyway. We now have a generation of small low-cost microcontrollers that have enough on-chip memory and processing power to support the use of an RTOS. Developing using this approach is therefore much more accessible.

Getting Started

In this chapter we will first look at setting up an introductory RTOS project for a Cortex-M-based microcontroller. Next, we will go through each of the RTOS primitives and how they influence the design of our application code. Finally, when we have a clear understanding of the RTOS features, we will take a closer look at the RTOS configuration file.

Setting Up a Project

The first exercise in the examples accompanying this book provides a PDF document giving a detailed step-by-step guide for setting up an RTOS project. Here we will look at the main differences between a standard C program and an RTOS-based program. First, our μVision project is defined in the normal way. This means that we start a new project and select a microcontroller from the component database. This will add the startup code and configure the compiler, linker, simulation model, and JTAG programming algorithms. Next, we add an empty C module and save it as main.c to start a C-based application. This will give us a project structure similar to that shown below.

Figure 6.1
A minimal application program consists of an assembler file for the startup code and a C module.

To make this into an RTOS project we must add an RTX configuration file and the RTOS library:

Figure 6.2
The RTOS configuration is held in the file RTX_Config_CM.c which must be added to your project.

As its name implies, this file holds the configuration settings for the RTOS. The default version of this file can be found in C:\Keil\ARM\Startup\, when you are using the default installation path.

We will examine this file in more detail later, after we have looked more closely at the RTOS and understand what needs to be configured.

To enable our C code to access the CMSIS RTOS API, we need to add an include file to all our application files that use RTOS functions. To do this you must add the following include file in main.c:

```
#include <cmsis_os.h>
```

We must let the debugger know that we are using the RTOS so it can provide additional debug support. This is done by selecting RTX Kernel in the Options for Target\Target menu.

Figure 6.3
Additional debug support is added to the project by enabling the OS support in the Options for Target menu.

First Steps with CMSIS RTOS

The RTOS itself consists of a scheduler, which supports round-robin, preemptive and cooperative multitasking of program threads, as well as time and memory management services. Interthread communication is supported by additional RTOS objects, including signal triggering, semaphores, mutex, and a mailbox system. As we will see, interrupt handling can also be accomplished by prioritized threads that are scheduled by the RTOS kernel.

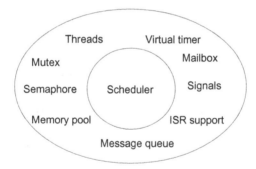

Figure 6.4
The RTOS kernel contains a scheduler that runs program code as threads. Communication between threads is accomplished by RTOS objects such as events, semaphores, mutexes, and mailboxes. Additional RTOS services include time and memory management and interrupt support.

Threads

The building blocks of a typical C program are functions that we call to perform a specific procedure and that then return to the calling function. In CMSIS RTOS the basic unit of execution is a "thread." A thread is very similar to a C procedure but has some very fundamental differences.

```
unsigned int procedure (void)        void thread (void)
{                                     {
     ......                               while(1)
   return(ch);                            {
}                                            ......
                                         }
                                      }
```

While we always return from our C function, once started, an RTOS thread must contain a loop so that it never terminates and thus runs forever. You can think of a thread as a mini self-contained program that runs within the RTOS.

An RTOS program is made up of a number of threads, which are controlled by the RTOS scheduler. This scheduler uses the systick timer to generate a periodic interrupt as a time base. The scheduler will allot a certain amount of execution time to each thread. So thread1 will run for 100 ms then be descheduled to allow thread2 to run for a similar period; thread2 will give way to thread3; and finally control passes back to thread1. By allocating these slices of runtime to each thread in a round-robin fashion, we get the appearance of all three threads running in parallel to each other.

Conceptually we can think of each thread as performing a specific functional unit of our program with all threads running simultaneously. This leads us to a more object-orientated design, where each functional block can be coded and tested in isolation and then integrated into a fully running program. This not only imposes a structure on the design of our final application but also aids debugging, as a particular bug can be easily isolated to a specific thread. It also aids code reuse in later projects. When a thread is created, it is also allocated its own thread ID. This is a variable that acts as a handle for each thread and is used when we want to manage the activity of the thread.

```
osThreadId id1,id2,id3;
```

In order to make the thread-switching process happen, we have the code overhead of the RTOS and we have to dedicate a CPU hardware timer to provide the RTOS time reference. In addition, each time we switch running threads, we have to save the state of all the thread variables to a thread stack. Also, all the runtime information about a thread is stored in a thread control block, which is managed by the RTOS kernel. Thus the "context switch time," that is, the time to save the current thread state and load up and start the next thread, is a crucial figure and will depend on both the RTOS kernel and the design of the underlying hardware.

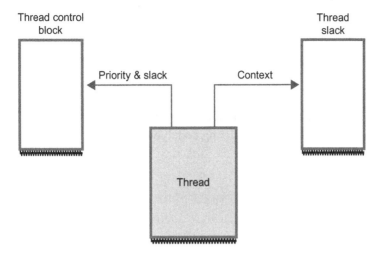

Figure 6.5
Each thread has its own stack for saving its data during a context switch. The thread control block is used by the kernel to manage the active thread.

The thread control block contains information about the status of a thread. Part of this information is its run state. In a given system, only one thread will be in the running state and all the others will be suspended but ready to run or in a wait state, waiting to be triggered by an OS event. The RTOS has various methods of inter-thread communication (i.e., signals, semaphores, messages, etc.). Here a thread may be suspended to wait to be signaled by another thread or interrupt before it resumes its ready state, whereupon it can be placed into the running state by the RTOS scheduler.

Table 6.1: At Any Given Moment a Single Thread May Be Running

Running	The currently running thread
Ready	Threads ready to run
Wait	Blocked threads waiting for an OS event

The remaining threads will be ready to run and will be scheduled by the kernel. Threads may also be waiting pending an OS event. When this occurs they will return to the ready state and be scheduled by the kernel.

Starting the RTOS

To build a simple RTOS program we declare each thread as a standard C function and also declare a thread ID variable for each function.

```
void thread1 (void);
void thread2 (void);
osThreadId thrdID1, thrdID2;
```

By default the CMSIS RTOS scheduler will start running when main() is entered and the main() function becomes the first active thread. Once in main() we can create further threads. It is also possible to configure CMSIS RTOS not to start automatically. The cmsis_os.h include file contains a define:

```
#define osFeature_MainThread 1
```

If this is changed to 0 then the main() function does not become a thread and the RTOS must be started explicitly. You can run any initializing code you want, before starting the RTOS.

```
void main (void)
{
  IODIR1 = 0x00FF0000;      / Do any C code you want
  osKernelStart(osThreadDef(Thread1),NULL);      //Start the RTOS
}
```

The osKernelStart function launches the RTOS but only starts the first thread running. After the OS has been initialized, control will be passed to this thread. When the first thread is created it is also assigned a priority. If there are a number of threads ready to run and they all have the same priority, they will be allotted runtime in a round-robin fashion. However, if a thread with a higher priority becomes ready to run, the RTOS scheduler will deschedule the currently running thread and start the high-priority thread running. This is called preemptive priority-based scheduling. When assigning priorities you have to be careful because the high-priority thread will continue to run until it enters a waiting state or until a thread of equal or higher priority is ready to run.

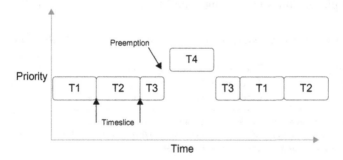

Figure 6.6
Threads of equal priority will be scheduled in a round-robin fashion. High-priority threads will preempt low-priority threads and enter the running state "on demand."

It is also possible to detect if the RTOS is running with osKernelRunning().

```
void main (void)
{
  if(!osKernelRunning()){
```

```
    osKernelStart(osThreadDef(Thread1),NULL);  //Start the RTOS
    }else{
while(1){
  main_thread();   //main is a thread so do stuff here
      }
    }
  }
```

This allows us to enter main() and detect if the kernel has started automatically or needs to be started explicitly.

Exercise: A First CMSIS RTOS Project

This project will take you through the steps necessary to convert a standard C application to a CMSIS RTOS application.

Open the project in C:\exercises\CMSIS RTOS first project.

```
#include "stm32f10x.h"
int main (void) {
  for (;;);
}
```

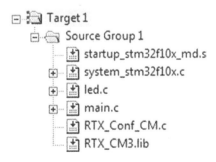

Here we have a simple program that runs to main and sits in a loop forever.

Right click on the Source Group 1 folder and add the RTX_Conf_CM.c file and the RTX library. You will need to change the "types of file" filter to see the library file.

Now your project should look like this.

In main.c add the CMSIS RTOS header file.

```
#include "cmsis_os.h"
```

In the Options for Target\Target tab set RTX Kernel as the OS.

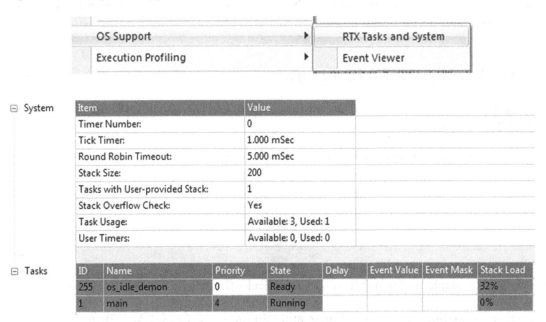

Operating system: | RTX Kernel ▼

Build the project and start the debugger.

The code will reach the for loop in main() and halt.

Open the Debug\OS Support\RTX Tasks and System window.

| OS Support ▶ | RTX Tasks and System |
| Execution Profiling ▶ | Event Viewer |

System

Item	Value
Timer Number:	0
Tick Timer:	1.000 mSec
Round Robin Timeout:	5.000 mSec
Stack Size:	200
Tasks with User-provided Stack:	1
Stack Overflow Check:	Yes
Task Usage:	Available: 3, Used: 1
User Timers:	Available: 0, Used: 0

Tasks

ID	Name	Priority	State	Delay	Event Value	Event Mask	Stack Load
255	os_idle_demon	0	Ready				32%
1	main	4	Running				0%

This is a new debug window that shows us the status of the RTOS and status of the running threads.

While this project does not actually do anything, it demonstrates the few steps necessary to start using CMSIS RTOS.

Creating Threads

Once the RTOS is running, there are a number of system calls that are used to manage and control the active threads. When the first thread is launched it is not assigned a thread ID. The first RTOS function we must therefore call is osThreadGetId() which returns the thread ID number. This is then stored in its ID handle "thread1." When we want to refer to this thread in future OS calls, we use this handle rather than the function name of the thread.

```
osThreadId main_id;      //create the thread handle
void main (void)
{
/* Read the Thread-ID of the main thread */
  main_id = osThreadGetId ();
  while(1)
  {
    .........
  }
}
```

Once we have obtained the thread number, we can use the main thread to create further active threads. This is done in two stages. First a thread structure is defined; this allows us to define the thread operating parameters. Then the thread is created from within the main thread.

```
osThreadId main_id;  //create the thread handles
osThreadId thread1_id;
void thread1 (void const *argument);   //function prototype for thread1
osThreadDef(thread1, osPriorityNormal, 1, 0);  //thread definition structure
void main (void)
{
/* Read the Thread-ID of the main thread */
    main_id = osThreadGetId ();
/* Create the second thread and assign its priority */
    Thread1_id = osThreadCreate(osThread(thread1), NULL);
    while(1)
    {
      .........
    }
}
```

The thread structure requires us to define the start address of the thread function, its thread priority, the number of instances of the thread that will be created, and its stack size. We will look at these parameters in more detail later. Once the thread structure has been defined the thread can be created using the osThreadCreate() API call. This creates the thread and starts it running. It is also possible to pass a parameter to the thread when it starts.

```
Thread1_id = osThreadCreate(osThread(thread1), startupParameter);
```

When each thread is created, it is also assigned its own stack for storing data during the context switch. This should not be confused with the Cortex processor stack; it is really a block of memory that is allocated to the thread. A default stack size is defined in the RTOS configuration file (we will see this later) and this amount of memory will be allocated to each thread unless we override it. If necessary a thread can be given additional memory resources by defining a bigger stack size in the thread structure.

```
osThreadDef(thread1, osPriorityNormal, 1, 1024);   //thread definition structure
```

Exercise: Creating and Managing Threads

In this project, we will create and manage some additional threads.

Open the project in c:\exercises\CMSIS RTOS Threads.

When the RTOS starts main() runs as a thread and in addition we will create two additional threads. First, we will create handles for each of the threads and then define the parameters of each thread. These include the priority the thread will run at, the number of instances of each thread we will create, and its stack size (the amount of memory allocated to it); zero indicates it will have the default stack size.

```
osThreadId main_ID,led_ID1,led_ID2;
osThreadDef(led_thread2, osPriorityAboveNormal, 1, 0);
osThreadDef(led_thread1, osPriorityNormal, 1, 0);
```

Then in the main() function the two threads are created.

```
led_ID2 =  osThreadCreate(osThread(led_thread2), NULL);
led_ID1 =  osThreadCreate(osThread(led_thread1), NULL);
```

When the thread is created we can pass it a parameter in place of the NULL define.

Build the project and start the debugger.

Start running the code and open the Debug\OS Support\RTX Tasks and System window.

ID	Name	Priority	State	Delay	Event Value	Event Mask	Stack Load
255	os_idle_demon	0	Ready				32%
3	led_thread1	5	Running				0%
2	led_thread2	5	Ready				32%
1	main	5	Ready				32%

Now we have four active threads with one running and the others ready.

Open the Debug\OS Support\Event Viewer window.

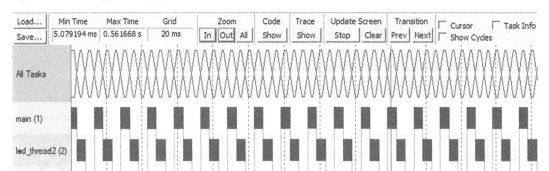

The event viewer shows the execution of each thread as a trace against time. This allows you to visualize the activity of each thread and get a feel for the amount of CPU time consumed by each thread.

Now open the Peripherals\General Purpose IO\GPIOB window.

Our two LED threads are each toggling a GPIO port pin. Leave the code running and watch the pins toggle for a few seconds.

```
void led_thread2 (void const *argument) {
  for (;;) {
  LED_On(1);
  delay(500);
  LED_Off(1);
  delay(500);
  }}
```

Each thread calls functions to switch an LED on and off and uses a delay function between each on and off. Several important things are happening here. First, the delay function can be safely called by each thread. Each thread keeps local variables in its stack so they cannot be corrupted by any other thread. Second, none of the threads blocks; each one runs for its full allocated timeslice, mostly sitting in the delay loop wasting cycles. Finally, there is no synchronization between the threads. They are running as separate "programs" on the CPU and as we can see from the GPIO debug window the toggled pins appear random.

Thread Management and Priority

When a thread is created it is assigned a priority level. The RTOS scheduler uses a thread's priority to decide which thread should be scheduled to run. If a number of threads are ready to run, the thread with the highest priority will be placed in the run state. If a high-priority thread becomes ready to run, it will preempt a running thread of lower priority. Importantly a high-priority thread running on the CPU will not stop running unless it blocks on an RTOS API call or is preempted by a higher priority thread. A thread's priority is defined in the thread structure and the following priority definitions are available. The default priority is osPriorityNormal.

Table 6.2: CMSIS Priority Levels

osPriorityIdle
osPriorityLow
osPriorityBelowNormal
osPriorityNormal
osPriorityAboveNormal
osPriorityHigh
osPriorityRealTime
osPriorityError

Threads with the same priority level will use round-robin scheduling. Higher priority threads will preempt lower priority threads.

Once the threads are running, there are a small number of OS system calls that are used to manage the running threads. It is also then possible to elevate or lower a thread's priority either from another function or from within its own code.

```
osStatus  osThreadSetPriority(threadID, priority);
osPriority  osThreadGetPriority(threadID);
```

As well as creating threads, it is also possible for a thread to delete itself or another active thread from the RTOS. Again we use the thread ID rather than the function name of the thread.

```
osStatus = osThreadTerminate (threadID1);
```

Finally, there is a special case of thread switching where the running thread passes control to the next ready thread of the same priority. This is used to implement a third form of scheduling called cooperative thread switching.

```
osStatus osThreadYeild();//switch to next ready to run thread
```

Exercise: Creating and Managing Threads II

In this exercise we will look at assigning different priorities to threads and also how to create and terminate threads dynamically.

Open the project in c:\exercises\CMSIS RTOS threads.

Change the priority of each LED thread to AboveNormal.

```
osThreadDef(led_task2, osPriorityAboveNormal, 1, 0);
osThreadDef(led_task1, osPriorityAboveNormal, 1, 0);
```

Build the project and start the debugger.

Start running the code.

Open the Debug\OS support\Event Viewer window.

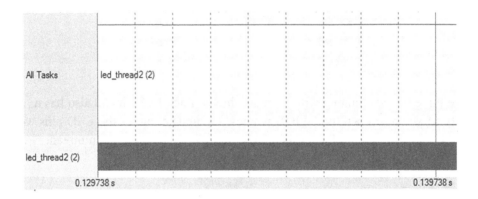

Here we can see thread2 running but no sign of thread1. Looking at the main() function and its coverage monitor shows us what has gone wrong.

```
55 ⊟int main (void) {
56 │    LED_Init ();
57 │    led_ID2 = osThreadCreate(osThread(led_task2), NULL);
58 │    led_ID1 = osThreadCreate(osThread(led_task1), NULL);
```

Executed ⟶
Not executed ⟋

Main is running at normal priority and created an LED thread at a higher priority. This thread immediately preempted main and stopped the creation of the second LED thread. To make it even worse, the LED thread never blocks so it will run forever preventing the main thread from ever running.

Exit the debugger and add the following code in the main() function before the osThreadCreate API calls.

```
main_ID = osThreadGetId ();
osThreadSetPriority(main_ID,osPriorityAboveNormal);
```

First, we set the main thread handle and then raise the priority of main to the same level as the LED threads.

Since we are only using main to create the LED threads we can stop running it by adding the following line of code after the osThreadCreate() functions.

```
osThreadTerminate(main_ID);
```

Build the code and start the debugger.

Open the RTX Tasks and System window.

ID	Name	Priority	State
255	os_idle_demon	0	Ready
3	led_thread1	5	Ready
2	led_thread2	5	Running

Now we have three threads running with no main thread. Each LED thread also has a higher priority. If you also open the GPIOB window the update rate of the LED pins would have changed because we are no longer using any runtime to sit in the main thread.

Multiple Instances

One of the interesting possibilities of an RTOS is that you can create multiple running instances of the same base thread code. So, for example, you could write a thread to control a UART and then create two running instances of the same thread code. Here each instance of the UART code could manage a different UART.

First, we can create the thread structure and set the number of thread instances to two.

```
osThreadDef(thread1, osPriorityNormal, 2, 0);
```

Then we can create two instances of the thread assigned to different thread handles. A parameter is also passed to allow each instance to identify which UART it is responsible for.

```
ThreadID_1_0 = osThreadCreate(osThread(thread1), UART1);
ThreadID_1_1 = osThreadCreate(osThread(thread1), UART2);
```

Exercise: Multiple Thread Instances

In this project, we will look at creating one thread and then create multiple runtime instances of the same thread.

Open the project in c:\exercises\CMSIS RTOS multiple instance.

This project performs the same function as the previous LED flasher program. However, we now have one LED switcher function that uses an argument passed as a parameter to decide which LED to flash.

```
void ledSwitcher (void const *argument) {
    for (;;) {
    LED_On((uint32_t)argument);
    delay(500);
    LED_Off((uint32_t)argument);
    delay(500);
  }
}
```

When we define the thread we adjust the instances parameter to 2.

```
osThreadDef(ledSwitcher, osPriorityNormal, 2, 0);
```

Then, in the main thread, we can create two threads that are different instances of the same base code. We pass a different parameter that corresponds to the LED that will be toggled by the instances of the thread.

```
led_ID1 =  osThreadCreate(osThread(ledSwitcher),(void *) 1UL);
led_ID2 =  osThreadCreate(osThread(ledSwitcher),(void *) 2UL);
```

Build the Code and Start the Debugger

Start running the code and open the RTX Tasks and System window.

ID	Name	Priority	State
255	os_idle_demon	0	Ready
3	ledSwitcher	4	Ready
2	ledSwitcher	4	Running

Here we can see both instances of the ledSwitcher() thread each with a different ID.

Examine the Call Stack + Locals window.

⊟ ◆ ledSwitcher : 2	0x08000318	
⊞ ◆ delay	0x08000310	
⊟ ◆ ledSwitcher	0x0800032A	
⊞ ◆ argument	0x00000001	
⊟ ◆ ledSwitcher : 3	0x08000318	
⊞ ◆ delay	0x08000314	
⊟ ◆ ledSwitcher	0x08000338	
⊞ ◆ argument	0x00000002	

Here we can see both instances of the ledSwitcher() threads and the state of their variables. A different argument has been passed to each instance of the thread.

Time Management

As well as running your application code as threads, the RTOS also provides some timing services that can be accessed through RTOS system calls.

Time Delay

The most basic of these timing services is a simple timer delay function. This is an easy way of providing timing delays within your application. Although the RTOS kernel size is quoted as 5 KB, features such as delay loops and simple scheduling loops are often part of a non-RTOS application and would consume code bytes anyway, so the overhead of the RTOS can be less than it immediately appears.

```
void osDelay (uint32_t millisec)
```

This call will place the calling thread into the WAIT_DELAY state for the specified number of milliseconds. The scheduler will pass execution to the next thread in the READY state.

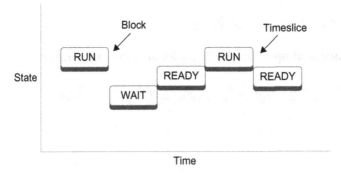

Figure 6.7

During their lifetime thread move through many states. Here a running thread is blocked by an osDelay() call so it enters a wait state. When the delay expires, it moves to ready. The scheduler will place it in the run state. If its timeslice expires, it will move back to ready.

When the timer expires, the thread will leave the wait_delay state and move to the READY state. The thread will resume running when the scheduler moves it to the RUNNING state. If the thread then continues executing without any further blocking of OS calls, it will be descheduled at the end of its timeslice and be placed in the ready state, assuming another thread of the same priority is ready to run.

Waiting for an Event

In addition to a pure time delay, it is possible to make a thread halt and enter the waiting state until the thread is triggered by another RTOS event. RTOS events can be a signal, message, or mail event. The osWait() API call also has a timeout period defined in milliseconds that allows the thread to wake up and continue execution if no event occurs.

```
osStatus osWait (uint32_t millisec)
```

When the interval expires, the thread moves from the wait to the READY state and will be placed into the running state by the scheduler. We will use this function later when we look at the RTOS thread intercommunication methods.

Exercise: Time Management

In this exercise we will look at using the basic time delay function.

Open the project in C:\exercises\CMSIS RTOS time management.

This is our original LED flasher program but the simple delay function has been replaced by the osDelay API call. LED2 is toggled every 100 ms and LED1 is toggled every 500 ms.

```
void ledOn (void const *argument) {
    for (;;) {
    LED_On(1);
  osDelay(500);
  LED_Off(1);
osDelay(500);
}}
```

Build the project and start the debugger.

Start running the code and open the event viewer window.

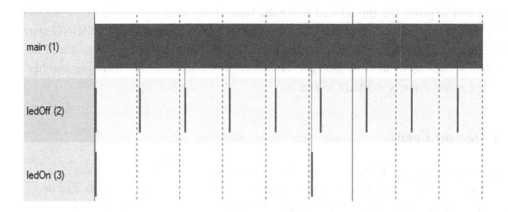

Now we can see that the activity of the code is very different. When each of the LED thread reaches the osDelay API call it "blocks" and moves to a waiting state. The main thread will be in a ready state so the scheduler will start running it. When the delay period has timed out, the LED threads will move to the ready state and will be placed into the running state by the scheduler. This gives us a multithreaded program where CPU runtime is efficiently shared between threads.

Virtual Timers

The CMSIS RTOS API can be used to define any number of virtual timers, which act as countdown timers. When they expire, they will run a user callback function to perform a specific action. Each timer can be configured as a one shot or repeat timer. A virtual timer is created by first defining a timer structure.

```
osTimerDef(timer0,led_function);
```

This defines a name for the timer and the name of the callback function. The timer must then be instantiated in an RTOS thread.

```
osTimerId timer0_handle = ocTimerCreate (timer(timer0), osTimerPeriodic, (void *)0);
```

This creates the timer and defines it as a periodic timer or a single shot timer (osTimerOnce). The final parameter passes an argument to the callback function when the timer expires.

```
osTimerStart (timer0_handle,0x100);
```

The timer can then be started at any point in a thread; the timer start function invokes the timer by its handle and defines a count period in milliseconds.

Exercise: Virtual Timer

In this exercise, we will configure a number of virtual timers to trigger a callback function at various frequencies.

Open the project in c:\exercises\CMSIS RTOS timer.

This is our original LED flasher program and code has been added to create four virtual timers to trigger a callback function. Depending on which timer has expired, this function will toggle an additional LED.

The timers are defined at the start of the code.

```
osTimerDef(timer0_handle, callback);
osTimerDef(timer1_handle, callback);
osTimerDef(timer2_handle, callback);
osTimerDef(timer3_handle, callback);
```

They are then initialized in the main() function.

```
osTimerId timer0 = osTimerCreate(osTimer(timer0_handle), osTimerPeriodic, (void *)0);
osTimerId timer1 = osTimerCreate(osTimer(timer1_handle), osTimerPeriodic, (void *)1);
osTimerId timer2 = osTimerCreate(osTimer(timer2_handle), osTimerPeriodic, (void *)2);
osTimerId timer3 = osTimerCreate(osTimer(timer3_handle), osTimerPeriodic, (void *)3);
```

Each timer has a different handle and ID and passes a different parameter to the common callback function.

```
void callback(void const *param){
switch((uint32_t) param){
case 0:
  GPIOB->ODR ^= 0x8;
break;
case 1:
  GPIOB->ODR ^= 0x4;
break;
case 2:
  GPIOB->ODR ^= 0x2;
break;
case 3:
break;
}
```

When triggered, the callback function uses the passed parameter as an index to toggle the desired LED.

In addition to configuring the virtual timers in the source code, the timer thread must be enabled in the RTX configuration file.

Open the RTX_Conf_CM.c file and press the Configuration Wizard tab.

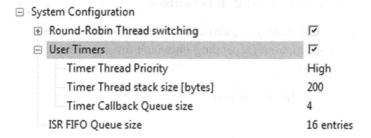

In the System Configuration section, make sure the User Timers box is ticked. If this thread is not created the timers will not work.

Build the project and start the debugger.

Run the code and observe the activity of the GPIOB pins in the peripheral window.

There will also be an additional thread running in the RTX threads and System window.

ID	Name	Priority	State	Delay
255	os_idle_demon	0	Ready	
4	ledOn	4	Wait_DLY	29
3	ledOff	4	Wait_AND	
2	osTimerThread	6	Wait_MBX	
1	main	4	Running	

The osDelay() function provides a relative delay from the point at which the delay is started. The virtual timers provide an absolute delay that allows you to schedule code to run at fixed intervals.

Idle Demon

The final timer service provided by the RTOS is not really a timer, but this is probably the best place to discuss it. If during our RTOS program we have no thread running and no thread ready to run (e.g., they are all waiting on delay functions) then the RTOS will use the spare runtime to call an "idle demon" that is again located in the RTX_Config.c file. This idle code is in effect a low-priority thread within the RTOS that only runs when nothing else is ready.

```
void os_idle_demon (void)
  {
    for (;;) {
    /* HERE: include here optional user code to be executed when no thread runs.  */
    }
    } /* end of os_idle_demon */
```

You can add any code to this thread, but it has to obey the same rules as user threads. The idle demon is an ideal place to add power management routines.

Exercise Idle Thread

Open the project in c:\exercises\CMSIS RTOS idle.

This is a copy of the virtual timer project. Some code has been added to the idle thread that will execute when our process thread are descheduled.

Open the RTX_Conf_CM.c file and click the text editor tab.

Locate the os_idle_demon thread.

```
void os_idle_demon (void) {
int32_t i;
for (;;) {
GPIOB->ODR ^= 0x1;
for(i = 0;i < 0x500;i++);
}}
```

When we enter the idle demon it will toggle the least significant port bit.

Build the code and start the debugger.

Run the code and observe the state of the port pins in the GPIOB peripheral window.

This is a simple program that spends most of its time in the idle demon, so this code will be run almost continuously.

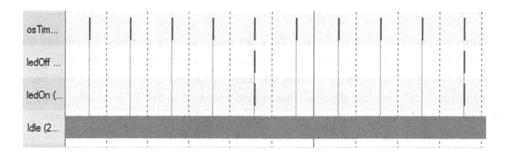

You can also see the activity of the idle demon in the event viewer. In a real project, the amount of time spent in the idle demon is an indication of spare CPU cycles.

The simplest use of the idle demon is to place the microcontroller into a low-power mode when it is not doing anything.

```
void os_idle_demon (void) {
__wfi();
}}
```

What happens next depends on the power mode selected in the microcontroller. At a minimum the CPU will halt until an interrupt is generated by the systick timer and execution of the scheduler will resume. If there is a thread ready to run then execution of the application code will resume. Otherwise, the idle demon will be reentered and the system will go back to sleep.

Interthread Communication

So far we have seen how application code can be defined as independent threads and how we can access the timing services provided by the RTOS. In a real application we need to be able to communicate between threads in order to make an application useful. To this end, a typical RTOS supports several different communication objects that can be used to link the threads together to form a meaningful program. The CMSIS RTOS API supports interthread communication with signals, semaphores, mutexes, mailboxes, and message queues.

Signals

When each thread is first created it has eight signal flags. These are stored in the thread control block. It is possible to halt the execution of a thread until a particular signal flag or group of signal flags are set by another thread in the system.

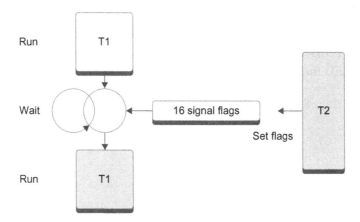

Figure 6.8
Each thread has eight signal flags. A thread may be placed into a waiting state until a pattern of flags is set by another thread. When this happens it will return to the ready state and wait to be scheduled by the kernel.

The signal wait system calls will suspend execution of the thread and place it into the wait_event state. Execution of the thread will not start until all the flags set in the signal wait API call have been set. It is also possible to define a periodic timeout, after which the waiting thread will move back to the ready state, so that it can resume execution when selected by the scheduler. A value of 0xFFFF defines an infinite timeout period.

```
osEvent osSignalWait (int32_t signals,uint32_t millisec);
```

If the signals variable is set to zero when osSignalWait is called then setting any flag will cause the thread to resume execution. Any thread can set or clear a signal on any other thread.

```
int32_t osSignalSet (osThreadId thread_id, int32_t signals);
int32_t osSignalClear (osThreadId thread_id, int32_t signals);
```

It is also possible for threads to read the state of their own or other threads' signal flags.

```
Int32_t osSignalGet (osThreadId threaded);
```

While the number of signals available to each thread defaults to eight, it is possible to enable up to 32 flags per thread by changing the signals defined in the CMSIS_os.h header file.

```
#define osFeature_Signals 16
```

Exercise: Signals

In this exercise we will look at using signals to trigger activity between two threads. While this is a simple program it introduces the concept of synchronizing the activity of threads together.

Open the project in c:\exercises\CMSIS RTOS signals.

This is a modified version of the LED flasher program; one of the threads calls the same LED function and uses osDelay() to pause the thread. In addition it sets a signal flag to wake up the second LED thread.

```
void led_Thread2 (void const *argument) {
for (;;) {
  LED_On(2);
  osSignalSet  (T_led_ID1,0x01);
  osDelay(500);
  LED_Off(2);
  osSignalSet  (T_led_ID1,0x01);
  osDelay(500);
}}
```

The second LED function waits for the signal flags to be set before calling the LED functions.

```
void led_Thread1 (void const *argument) {
for (;;) {
osSignalWait  (0x01,osWaitForever);
LED_On(1);
osSignalWait  (0x01,osWaitForever);
LED_Off(1);
}}
```

Build the project and start the debugger.

Open the GPIOB peripheral window and start running the code.

Now the port pins will appear to be switching on and off together. Synchronizing the threads gives the illusion that both threads are running in parallel.

This is a simple exercise but it illustrates the key concept of synchronizing activity between threads in an RTOS-based application.

RTOS Interrupt Handling

The use of signal flags is a simple and efficient method of triggering actions between threads running within the RTOS. Signal flags are also an important method of triggering RTOS threads from interrupt sources within the Cortex microcontroller. While it is possible to run C code in an ISR, this is not desirable within an RTOS if the interrupt code is going to run for more than a short period of time. This delays the timer tick and disrupts the

RTOS kernel. The systick timer runs at the lowest priority within the NVIC so there is no overhead in reaching the interrupt routine.

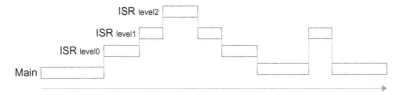

Figure 6.9
A traditional nested interrupt scheme supports prioritized interrupt handling, but has unpredictable stack requirements.

With an RTOS application, it is best to design the interrupt service code as a thread within the RTOS and assign it a high priority. The first line of code in the interrupt thread should make it wait for a signal flag. When an interrupt occurs, the ISR simply sets the signal flag and terminates. This schedules the interrupt thread, which services the interrupt and then goes back to waiting for the next signal flag to be set.

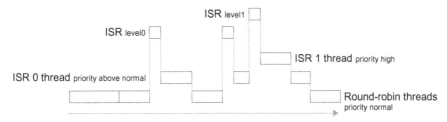

Figure 6.10
Within the RTOS, interrupt code is run as threads. The interrupt handlers signal the threads when an interrupt occurs. The thread priority level defines which thread gets scheduled by the kernel.

A typical interrupt thread will have the following structure:

```
void Thread3 (void)
{
while(1)
{
osSignalWait (isrSignal,waitForever);  // Wait for the ISR to trigger an event
.....  // Handle the interrupt
}  // Loop round and go back to sleep
}
```

The actual interrupt source will contain a minimal amount of code.

```
void IRQ_Handler (void)
{
osSignalSet (tsk3,isrSignal);  // Signal Thread 3 with an event
}
```

Exercise: Interrupt Signal

CMSIS RTOS does not introduce any latency in serving interrupts generated by user peripherals. However, operation of the RTOS may be disturbed if you lock out the systick interrupt for a long period of time. This exercise demonstrates a technique of signaling a thread from an interrupt and servicing the peripheral interrupt with a thread rather than a standard ISR.

Open the project in c:\exercises\CMSIS RTOS interrupt signal.

In the main() function we initialize the ADC and create an ADC thread that has a higher priority than all the other threads.

```
osThreadDef(adc_Thread, osPriorityAboveNormal, 1, 0);
int main (void) {
LED_Init ();
init_ADC ();
T_led_ID1 =  osThreadCreate(osThread(led_Thread1), NULL);
T_led_ID2 =  osThreadCreate(osThread(led_Thread2), NULL);
T_adc_ID =  osThreadCreate(osThread(adc_Thread), NULL);
```

However, there is a problem when we enter ()main: the code is running in unprivileged mode. This means that we cannot access the NVIC registers without causing a fault exception. There are several ways round this; the simplest is to give the threads privileged access by changing the setting in RTX_Conf_CM.c.

Thread Configuration	
Number of concurrent running threads	5
Default Thread stack size [bytes]	200
Main Thread stack size [bytes]	200
Number of threads with user-provided stack size	0
Total stack size [bytes] for threads with user-provid...	0
Check for stack overflow	☑
Processor mode for thread execution	Privileged mode

Here we have switched the thread execution mode to privileged, which gives the threads full access to the Cortex-M processor. As we have added a thread, we also need to increase the number of concurrent running threads.

Build the code and start the debugger.

Set breakpoints in LED_Thread2, ADC_Thread, and ADC1_2_IRQHandler.

```
  57        osDelay(500);
● 58        ADC1->CR2     |=   (1UL << 22);
  59        LED_Off(2);
```

```
  35   osSignalWait   ( 0x01,osWaitForever);
● 36   GPIOB->ODR = ADC1->DR;
```

```
  28  ⊟void ADC1_2_IRQHandler (void){
● 29   osSignalSet ( T_adc_ID,0x01);
```

Run the code.

You should hit the first breakpoint, which starts the ADC conversion. Then run the code again and you should enter the ADC interrupt handler. The handler sets the ADC_thread signal and quits. Setting the signal will cause the ADC thread to preempt any other running thread run the ADC service code, and then block waiting for the next signal.

Exercise: CMSIS RTX and SVC Exceptions

As we saw in the last example when we are in a thread it will be running in unprivileged mode. The simple solution is to allow threads to run in privileged mode, but this allows the threads full access to the Cortex-M processor potentially allowing runtime errors. In this exercise we will look at using the system call exception to enter privileged mode to run "system level" code.

Open the project in c:\exercises\CMSIS RTOS interrupt_SVC.

In the project we have added a new file called SVC_Tables.c. This is located in c:\Keil\ARM\RL\RTX\SRC\CM.

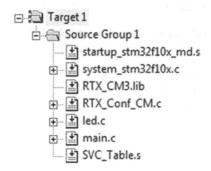

This is the lookup table for the SVC interrupts.

```
; Import user SVC functions here.
  IMPORT __SVC_1
  EXPORT SVC_Table
```

SVC_Table

```
; Insert user SVC functions here. SVC 0 used by RTX Kernel.
  DCD   __SVC_1    ; user SVC function
```

In this file we need to add import name and table entry for each __SVC function that we are going to use. In our example, we only need __SVC_1.

Now we can convert the ADC init function to a service call exception.

```
void __svc(1) init_ADC (void);
void __SVC_1 (void){
```

Build the project and start the debugger.

Step the code (F11) to the call to the init_ADC function and examine the operating mode in the register window. Here we are in thread mode, unprivileged, and using the process stack.

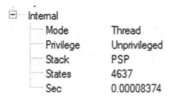

Now step into the function (F11) and step through the assembler until you reach the init_ADC C function.

```
⊟ ··· Intemal
       ···· Mode        Handler
       ···· Privilege   Privileged
       ···· Stack       MSP
       ···· States      4687
       ···· Sec         0.00008443
```

Now we are running in handler mode with privileged access and are using the MSP.

This allows us the set up the ADC and also access the NVIC.

Semaphores

Like events, semaphores are a method of synchronizing activity between two and more threads.
Put simply, a semaphore is a container that holds a number of tokens. As a thread executes, it
will reach an RTOS call to acquire a semaphore token. If the semaphore contains one or more
tokens, the thread will continue executing and the number of tokens in the semaphore will be
decremented by one. If there are currently no tokens in the semaphore, the thread will be
placed in a waiting state until a token becomes available. At any point in its execution a thread
may add a token to the semaphore causing its token count to increment by one.

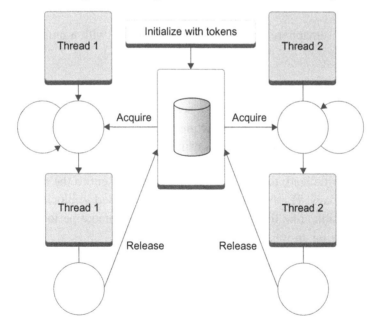

Figure 6.11
Semaphores are used to control access to program resources. Before a thread can access a
resource, it must acquire a token. If none is available, it waits. When it is finished with the
resource, it must return the token.

Figure 6.11 illustrates the use of a semaphore to synchronize two threads. First, the semaphore must be created and initialized with an initial token count. In this case, the semaphore is initialized with a single token. Both threads will run and reach a point in their code where they will attempt to acquire a token from the semaphore. The first thread to reach this point will acquire the token from the semaphore and continue execution. The second thread will also attempt to acquire a token, but as the semaphore is empty it will halt execution and be placed into a waiting state until a semaphore token is available.

Meanwhile, the executing thread can release a token back to the semaphore. When this happens, the waiting thread will acquire the token and leave the waiting state for the ready state. Once in the ready state the scheduler will place the thread into the run state so that thread execution can continue. While semaphores have a simple set of OS calls, they can be one of the more difficult OS objects to fully understand. In this section, we will first look at how to add semaphores to an RTOS program and then go on to look at the most useful semaphore applications.

To use a semaphore in the CMSIS RTOS you must first declare a semaphore container.

```
osSemaphoreId sem1;
osSemaphoreDef(sem1);
```

Then within a thread the semaphore container can be initialized with a number of tokens.

```
sem1 = osSemaphoreCreate(osSemaphore(sem1), SIX_TOKENS);
```

It is important to understand that semaphore tokens may also be created and destroyed as threads run. So, for example, you can initialize a semaphore with zero tokens and then use one thread to create tokens for the semaphore while another thread removes them. This allows you to design threads as producer and consumer threads.

Once the semaphore is initialized, tokens may be acquired and sent to the semaphore in a similar fashion to event flags. The osSemaphoreWait() call is used to block a thread until a semaphore token is available, like the osSignalWait() call. A timeout period may also be specified with 0xFFFF being an infinite wait.

```
osStatus osSemaphoreWait(osSemaphoreId semaphore_id, uint32_t millisec);
```

Once the thread has finished using the semaphore resource, it can send a token to the semaphore container.

```
osStatus osSemaphoreRelease(osSemaphoreId semaphore_id);
```

Exercise: Semaphore Signaling

In this exercise we will look at the configuration of a semaphore and use it to signal between two threads.

Open the exercise in C:\exercises\CMSIS RTOS semaphore.

The code first creates a semaphore called sem1 and initializes it with zero tokens.

```
osSemaphoreId sem1;
osSemaphoreDef(sem1);
int main (void) {
sem1 = osSemaphoreCreate(osSemaphore(sem1), 0);
```

The first thread waits for a token to be sent to the semaphore.

```
void led_Thread1 (void const *argument) {
  for (;;) {
    osSemaphoreWait(sem1, osWaitForever);
    LED_On(1);
    osDelay(500);
      LED_Off(1);
  }
}
```

The second thread periodically sends a token to the semaphore.

```
void led_Thread2 (void const *argument) {
  for (;;) {
    LED_On(2);
      osSemaphoreRelease(sem1);
    osDelay(500);
      LED_Off(2);
      osDelay(500);
      }}
```

Build the project and start the debugger.

Set a breakpoint in the LED_thread2 thread.

```
● 44        osSemaphoreRelease(sem1);
```

Run the code and observe the state of the threads when the breakpoint is reached.

ID	Name	Priority	State
255	os_idle_demon	0	Ready
3	led_Thread1	5	Wait_SEM
2	led_Thread2	4	Running
1	main	4	Ready

Now LED_thread1 is blocked waiting to acquire a token from the semaphore. LED_thread1 has been created with a higher priority than LED_thread2 so as soon as a token is placed in the semaphore it will move to the ready state and preempt the lower priority thread and start running. When it reaches the osSemaphoreWait() call it will again block.

Now block step the code (F10) and observe the action of the threads and the semaphore.

Using Semaphores

Although semaphores have a simple set of OS calls, they have a wide range of synchronizing applications. This makes them perhaps the most challenging RTOS object to understand. In this section, we will look at the most common uses of semaphores. These are taken from *The Little Book of Semaphores* by Allen B. Downy. This book may be freely downloaded from the URL given in the bibliography at the end of this book.

Signaling

Synchronizing the execution of two threads is the simplest use of a semaphore:

```
osSemaphoreId sem1;
osSemaphoreDef(sem1);
void thread1 (void)
{
sem1 = osSemaphoreCreate(osSemaphore(sem1), 0);
while(1)
{
FuncA();
osSemaphoreRelease(sem1)
}
}
void task2 (void)
{
```

```
while(1)
{
osSemaphoreWait(sem1,osWaitForever)
FuncB();
}
}
```

In this case the semaphore is used to ensure that the code in FuncA() is executed before the code in FuncB().

Multiplex

A multiplex is used to limit the number of threads that can access a critical section of code. For example, this could be a routine that accesses memory resources and can only support a limited number of calls.

```
osSemaphoreId multiplex;
osSemaphoreDef(multiplex);
Void thread1 (void)
{
multiplex = osSemaphoreCreate(osSemaphore(multiplex), FIVE_TOKENS);
while(1)
{
osSemaphoreWait(multiplex,osWaitForever)
ProcessBuffer();
osSemaphoreRelease(multiplex)
}
}
```

In this example, we initialize the multiplex semaphore with five tokens. Before a thread can call the ProcessBuffer() function, it must acquire a semaphore token. Once the function has been completed, the token is sent back to the semaphore. If more than five threads are attempting to call ProcessBuffer(), the sixth must wait until a thread has finished with ProcessBuffer() and returns its token. Thus the multiplex semaphore ensures that a maximum of five threads can call the ProcessBuffer() function "simultaneously."

Exercise: Multiplex

In this exercise we will look at using a semaphore to control access to a function by creating a multiplex.

Open the project in c:\exercises\CMSIS RTOS semaphore multiplex.

The project creates a semaphore called semMultiplex that contains one token.

Next six instances of a thread containing a semaphore multiplex are created.

Build the code and start the debugger.

Open the Peripherals\General Purpose IO\GPIOB window.

Run the code and observe how the threads set the port pins.

As the code runs only one thread at a time can access the LED functions so only one port pin is set.

Exit the debugger and increase the number of tokens allocated to the semaphore when it is created.

```
semMultiplex = osSemaphoreCreate(osSemaphore(semMultiplex), 3);
```

Build the code and start the debugger.

Run the code and observe the GPIOB pins.

Now three threads can access the LED functions "concurrently."

Rendezvous

A more generalized form of semaphore signaling is a rendezvous. A rendezvous ensures that two threads reach a certain point of execution. Neither may continue until both have reached the rendezvous point.

```
osSemaphoreId arrived1,arrived2;
osSemaphoreDef(arrived1);
osSemaphoreDef(arrived2);
void thread1 (void){
Arrived1 = osSemaphoreCreate(osSemaphore(arrived1),ZERO_TOKENS);
Arrived2 = osSemaphoreCreate(osSemaphore(arrived2),ZERO_TOKENS);
while(1){
FuncA1();
osSemaphoreRelease(Arrived1);
osSemaphoreWait(Arrived2,osWaitForever);
FuncA2();
}}
void thread2 (void) {
while(1){
```

```
FuncB1();
osSemaphoreRelease(semArrived2);
    osSemaphoreWait
(semArrived1,osWaitForever);
FuncB2();
}}
```

In the above case, the two semaphores will ensure that both threads will rendezvous and then proceed to execute FuncA2() and FuncB2().

Exercise: Rendezvous

In this project we will create two threads and have to reach a semaphore rendezvous before running the LED functions.

Open the exercise in c:\exercises\CMSIS RTOS semaphore rendezvous.

Build the project and start the debugger.

Open the Peripherals\General Purpose IO\GPIOB window.

Run the code.

Initially the semaphore code in each of the LED threads is commented out. Since the threads are not synchronized the GPIO pins will toggle randomly.

Exit the debugger.

Uncomment the semaphore code in the LED threads.

Build the project and start the debugger.

Run the code and observe the activity of the pins in the GPIOB window.

Now the threads are synchronized by the semaphore and run the LED functions "concurrently."

Barrier Turnstile

Although a rendezvous is very useful for synchronizing the execution of code, it only works for two functions. A barrier is a more generalized form of rendezvous that works to synchronize multiple threads.

```
osSemaphoreId count,barrier;
osSemaphoreDef(counter);
osSemaphoreDef(barrier);
```

```
Unsigned int count;
Turnstile    =    osSemaphoreCreate(osSemaphore(Turnstile), 0);
Turnstile2   =    osSemaphoreCreate(osSemaphore(Turnstile2), 1);
Mutex        =    osSemaphoreCreate(osSemaphore(Mutex), 1);
while(1)
{
        osSemaphoreWait(Mutex,0xffff);    //Allow only one task at a time to run this code
        count = count + 1;                // Increment count
        if(count = = 5)          //When last section of code reaches this point run his code
{
            osSemaphoreWait (Turnstile2,0xffff); //Lock the exit turnstile
            osSemaphoreRelease(Turnstile); //Unlock the first turnstile
}
osSemaphoreRelease(Mutex);                        //Allow other tasks to access the turnstile
osSemaphoreWait(Turnstile,0xFFFF);            //Turnstile Gate
osSemaphoeRelease(Turnstile);
criticalCode();
}
```

In this code we use a global variable to count the number of threads that have arrived at the barrier. As each function arrives at the barrier it will wait until it can acquire a token from the counter semaphore. Once acquired, the count variable will be incremented by one. Once we have incremented the count variable, a token is sent to the counter semaphore so that other waiting threads can proceed. Next, the barrier code reads the count variable. If this is equal to the number of threads that are waiting to arrive at the barrier, we send a token to the barrier semaphore.

In the example above, we are synchronizing five threads. The first four threads will increment the count variable and then wait at the barrier semaphore. The fifth and last thread to arrive will increment the count variable and send a token to the barrier semaphore. This will allow it to immediately acquire a barrier semaphore token and continue execution. After passing through the barrier, it immediately sends another token to the barrier semaphore. This allows one of the other waiting threads to resume execution. This thread places another token in the barrier semaphore, which triggers another waiting thread and so on. This final section of the barrier code is called a turnstile because it allows one thread at a time to pass the barrier.

Exercise: Semaphore Barrier

In this exercise we will use semaphores to create a barrier to synchronize multiple threads.

Open the project in c:\exercises\CMSIS RTOS semaphore barrier.

Open the exercise in c:\exercises\semaphore rendezvous.

Build the project and start the debugger.

Open the Peripherals\General Purpose IO\GPIOB window.

Run the code.

Initially, the semaphore code in each of the threads is commented out. Since the threads are not synchronized the GPIO pins will toggle randomly like in the rendezvous example.

Exit the debugger.

Uncomment the semaphore code in the threads.

Build the project and start the debugger.

Run the code and observe the activity of the pins in the GPIOB window.

Now the threads are synchronized by the semaphore and run the LED functions "concurrently."

Semaphore Caveats

Semaphores are an extremely useful feature of any RTOS. However, semaphores can be misused. You must always remember that the number of tokens in a semaphore is not fixed. During the runtime of a program, semaphore tokens may be created and destroyed. Sometimes this is useful, but if your code depends on having a fixed number of tokens available to a semaphore you must be very careful to always return tokens back to it. You should also rule out the possibility of accidently creating additional new tokens.

Mutex

Mutex stands for "mutual exclusion." In reality a mutex is a specialized version of a semaphore. Like a semaphore, a mutex is a container for tokens. The difference is that a mutex can only contain one token which cannot be created or destroyed. The principal use of a mutex is to control access to a chip resource such as a peripheral. For this reason a mutex token is binary and bounded. Apart from this it really works in the same way as a semaphore. First of all, we must declare the mutex container and initialize the mutex:

```
osMutexId uart_mutex;
osMutexDef (uart_mutex);
```

Once declared the mutex must be created in a thread:

```
uart_mutex = osMutexCreate(osMutex(uart_mutex));
```

Then any thread needing to access the peripheral must first acquire the mutex token:

```
osMutexWait(osMutexId mutex_id,uint32_t millisec;
```

Finally, when we are finished with the peripheral the mutex must be released:

```
osMutexRelease(osMutexId mutex_id);
```

Mutex use is much more rigid than semaphore use, but is a much safer mechanism when controlling absolute access to underlying chip registers.

Exercise: Mutex

In this exercise, our program writes streams of characters to the microcontroller UART from different threads. We will declare and use a mutex to guarantee that each thread has exclusive access to the UART until it has finished writing its block of characters.

Open the project in c:\exercises\mutex.

This project declares two threads and both write blocks of characters to the UART. Initially, the mutex is commented out.

```
void uart_Thread1 (void const *argument) {
uint32_t i;
  for (;;) {
//osMutexWait(uart_mutex, osWaitForever);
for(i = 0;i < 10;i++) SendChar('1');
SendChar('\n');
SendChar('\r');
//osMutexRelease(uart_mutex);
    }}
```

In each thread, the code prints out the thread number at the end of each block of characters; it then prints the carriage return and new line characters.

Build the code and start the debugger.

Open the UART #1 console window with View\Serial Windows\UART #1.

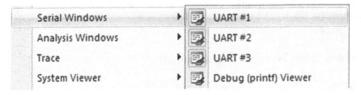

Start running the code and observe the output in the console window.

Here we can see that the output data stream is corrupted by each thread writing to the UART without any accessing control.

Exit the debugger.

Uncomment the mutex calls in each thread.

Build the code and start the debugger.

Observe the output of each thread in the console window.

Now the mutex guarantees each thread exclusive access to the UART while it writes each block of characters.

Mutex Caveats

Clearly, you must take care to return the mutex token when you are finished with the chip resource, or you will have effectively prevented any other thread from accessing it. You must also be extremely careful about using the osThreadTerminate() call on functions that control a mutex token. The Keil RTX RTOS is designed to be a small footprint RTOS so that it can run on even the very small Cortex microcontrollers. Consequently, there is no thread deletion safety. This means that if you delete a thread that is controlling a mutex token, you will destroy the mutex token and prevent any further access to the guarded peripheral.

Data Exchange

So far all of the interthread communication methods have only been used to trigger execution of threads; they do not support the exchange of program data between threads. Clearly in a real program we will need to move data between threads. This could be done by reading and writing to globally declared variables. In anything but a very simple program, trying to guarantee data integrity would be extremely difficult and prone to unforeseen errors. The exchange of data between threads needs a more formal asynchronous method of communication.

CMSIS RTOS provides two methods of data transfer between threads. The first method is a message queue, which creates a buffered data "pipe" between two threads. The message queue is designed to transfer integer values.

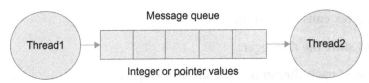

Figure 6.12
CMSIS RTOS supports a message queue that allows simple memory objects to be passed between threads.

The second form of data transfer is a mail queue. This is very similar to a message queue except that it transfers blocks of data rather than a single integer.

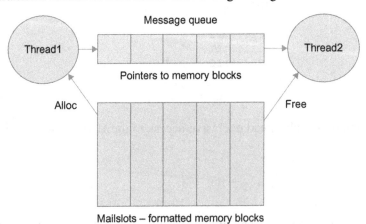

Figure 6.13
CMSIS RTOS also defines a mailbox system that allows complex memory objects to be transferred between threads. The mailbox is a combination of a memory pool and message queue.

Message Queue

To set up a message queue we first need to allocate the memory resources.

```
osMessageQId Q_LED;
osMessageQDef (Q_LED,16_Message_Slots,unsigned int);
```

This defines a message queue with 16 storage elements. In this particular queue, each element is defined as an unsigned integer. While we can post data directly into the message queue it is also possible to post a pointer to a data object.

```
osEvent result;
```

We also need to define an osEvent variable, which will be used to retrieve the queue data. The osEvent variable is a union that allows you to retrieve data from the message queue in a number of formats.

```
Union{
Uint32_t v
Void *p
Int32_t signals
}value
```

The osEvent union allows you to read the data posted to the message queue as an unsigned integer or a void pointer. Once the memory resources are created, we can declare the message queue in a thread.

```
Q_LED = osMessageCreate(osMessageQ(Q_LED),NULL);
```

Once the message queue has been created, we can put data into the queue from one thread.

```
osMessagePut(Q_LED,0x0,osWaitForever);
```

Then we can read from the queue in another.

```
result =  osMessageGet(Q_LED,osWaitForever);
LED_data = result.value.v;
```

Exercise: Message Queue

In this exercise, we will look at defining a message queue between two threads and then use it to send process data.

Open the project in c:\exercises\message queue.

```
osMessageQId Q_LED;
osMessageQDef (Q_LED,0x16,unsigned char);
```

```
osEvent result;
int main (void) {
LED_Init ();
Q_LED = osMessageCreate(osMessageQ(Q_LED),NULL);
```

We define and create the message queue in the main thread along with the event structure.

```
osMessagePut(Q_LED,0x1,osWaitForever);
osDelay(100);
```

Then in one of the threads we can post data and receive it in the second.

```
result =  osMessageGet(Q_LED,osWaitForever);
  LED_On(result.value.v);
```

Build the project and start the debugger.

Set a breakpoint in LED_thread1.

```
34 □void led_Thread1 (void const *argument) {
35 □   for (;;) {
36 |       result =  osMessageGet(Q_LED,osWaitForever);
●37 ||      LED_On(result.value.v);
```

Now run the code and observe the data as it arrives.

Memory Pool

While it is possible to post simple data values into the message queue, it is also possible to post a pointer to a more complex object. CMSIS RTOS supports dynamically allocatable memory in the form of a memory pool. Here we can declare a structure that combines a number of data elements.

```
typedef struct {
uint8_t LED0;
uint8_t LED1;
uint8_t LED2;
uint8_tLED3;
} memory_block_t;
```

Then we can create a pool of these objects as blocks of memory.

```
osPoolDef(led_pool,ten_blocks,memory_block_t);
osPoolId(led_pool);
```

Next we can create the memory pool by declaring it in a thread.

```
led_pool = osPoolCreate(osPool(led_pool));
```

Now we can allocate a memory pool within a thread as follows:

```
memory_block_t *led_data;
*led_data = (memory_block_t *) osPoolAlloc(led_pool);
```

Then we can populate it with data.

```
led_data->LED0 = 0;
led_data->LED1 = 1;
led_data->LED2 = 2;
led_data->LED3 = 3;
```

It is then possible to place the pointer to the memory block in a message queue so that the data can be accessed by another thread.

```
osMessagePut(Q_LED,(uint32_t)led_data,osWaitForever);
```

The data is then received in another thread as follows.

```
osEvent event; memory_block_t * received;
event = osMessageGet(Q_LED,osWatiForever);
*received = (memory_block *)event.value.p;
led_on(received->LED0);
```

Once the data in the memory block has been used the block must be released back to the memory pool for reuse.

```
osPoolFree(led_pool,received);
```

Mail Queue

While memory pools can be used as data buffers within a thread, CMSIS RTOS also implements a mail queue, which is a combination of a memory pool and message queue. The mail queue uses a memory pool to create formatted memory blocks and passes pointers to these blocks in a message queue. This allows the data to stay in an allocated memory block while we only move a pointer between the different threads. A simple mail queue API makes this easy to set up and use. First, we need to declare a structure for the mailslot similar to the one we used for the memory pool.

```
typedef struct {
  uint8_t LED0;
  uint8_t LED1;
  uint8_t LED2;
  uint8_t LED3;
} mail_format;
```

This message structure is the format of the memory block that is allocated in the mail queue. Now we can create the mail queue and define the number of memory block "slots" in the mail queue.

```
osMailQDef(mail_box, sixteen_mail_slots, mail_format);
osMailQId mail_box;
```

Once the memory requirements have been allocated, we can create the mail queue in a thread.

```
mail_box = osMailCreate(osMailQ(mail_box), NULL);
```

Once the mail queue has been instantiated, we can post a message. This is different from the message queue in that we must first allocate a mailslot and populate it with data.

```
mail_format *LEDtx;
LEDtx = (mail_format*)osMailAlloc(mail_box, osWaitForever);
```

First declare a pointer in the mailslot format and then allocate this to a mailslot. This locks the mailslot and prevents it from being allocated to any other thread. If all of the mailslots are in use, the thread will block and wait for a mailslot to become free. You can define a timeout in milliseconds that will allow the thread to continue if a mailslot has not become free.

Once a mailslot has been allocated, it can be populated with data and then posted to the mail queue.

```
LEDtx->LED0 = led0[index];
LEDtx->LED1 = led1[index];
LEDtx->LED2 = led2[index];
LEDtx->LED3 = led3[index];
osMailPut(mail_box, LEDtx);
```

The receiving thread must declare a pointer in the mailslot format and in an osEvent structure.

```
osEvent evt;
mail_format *LEDrx;
```

Then in the thread loop we can wait for a mail message to arrive.

```
evt = osMailGet(mail_box, osWaitForever);
```

We can then check the event structure to see if it is indeed a mail message and extract the data.

```
if (evt.status == osEventMail) {
  LEDrx = (mail_format*)evt.value.p;
```

Once the data in the mail message has been used, the mailslot must be released so it can be reused.

```
osMailFree(mail_box, LEDrx);
```

Exercise: Mailbox

This exercise demonstrates configuration of a mailbox and using it to post messages between threads.

Open the project in c:\exercises\CMSIS RTOS mailbox.

The project uses the LED structure used in the description above to define a 16-slot mailbox.

```
osMailQDef(mail_box, 16, mail_format);
osMailQId mail_box;
int main (void) {
  LED_Init();
  mail_box = osMailCreate(osMailQ(mail_box), NULL);
```

A producer thread then allocates a mailslot, fills it with data, and posts it to the mail queue. The receiving thread waits for a mail message to arrive then reads the data. Once the data has been used the mailslot is released.

Build the code and start the debugger.

Set a breakpoint in the consumer thread and run the code.

```
  52 |              evt = osMailGet(mail_box, osWaitForever);
● 53 ⊟|             if (evt.status == osEventMail) {
  54 |                  LEDrx = (mail_format*)evt.value.p;
```

Observe the mailbox messages arriving at the consumer thread.

Configuration

So far we have looked at the CMSIS RTOS API. This includes thread management functions, time management, and interthread communication. Now that we have a clear idea of exactly what the RTOS kernel is capable of, we can take a more detailed look at the configuration file. There is one configuration file for all of the Cortex-M processors and microcontrollers.

⊟ Thread Configuration	
Number of concurrent running threads	3
Default Thread stack size [bytes]	200
Main Thread stack size [bytes]	200
Number of threads with user-provided stack size	0
Total stack size [bytes] for threads with user-provided stack size	0
Check for stack overflow	☑
Processor mode for thread execution	Unprivileged mode
⊟ RTX Kernel Timer Tick Configuration	
Use Cortex-M SysTick timer as RTX Kernel Timer	☑
Timer clock value [Hz]	72000000
Timer tick value [us]	1000
⊟ System Configuration	
⊟ Round-Robin Thread switching	☑
Round-Robin Timeout [ticks]	5
⊟ User Timers	☑
Timer Thread Priority	High
Timer Thread stack size [bytes]	200
Timer Callback Queue size	4
ISR FIFO Queue size	16 entries

Figure 6.14
The CMSIS RTOS configuration values are held in the RTX_Config.c file. All of the RTOS
parameters can be configured through a microvision wizard.

Like the other configuration files, the RTX_Config_CM.c file is a template file that
presents all the necessary configurations as a set of menu options.

Thread Definition

In the thread definition section, we define the basic resources that will be required by the
CMSIS RTOS threads. For each thread, we allocate a defined stack space. (In the above
example this is 200 bytes.) We also define the maximum number of concurrently running
threads. Thus the amount of RAM required for the above example can easily be computed as
200×6 or 1200 bytes. If some of our threads need a larger stack space, then a larger stack can
be allocated when the thread is created. If we are defining custom stack sizes, we must define
the number of threads with custom stacks. Again, the RAM requirement is easily calculated.

During development, the CMSIS RTOS can trap stack overflows. When this option is
enabled, an overflow of a thread stack space will cause the RTOS kernel to call the
os_error function, which is located in the RTX_Conf_CM.c file. This function gets an error
code and then sits in an infinite loop. The stack checking option is intended for use during
debugging and should be disabled on the final application to minimize the kernel overhead.
However, it is possible to modify the os_error() function if enhanced error protection is
required in the final release. The final option in the thread definition section allows you to
define the number of user timers. It is a common mistake to leave this set at zero. If you do

not set this value to match the number of virtual timers in use by your application, the user timer API calls will fail to work. The thread definition section also allows us to select whether the threads are running in privileged or unprivileged mode.

System Timer Configuration

The default timer for use with CMSIS RTOS is the Cortex-M systick timer, which is present on all Cortex-M processors. The input to the systick timer will generally be the CPU clock. It is possible to use a different timer by unchecking the "use systick option." If you do this there are two function stubs in the RTX_Config_CM.c file that allow you to initialize the alternative timer and acknowledge its interrupt.

```
int os_tick_init (void) {
   return (-1); /* Return IRQ number of timer (0..239) */
}
void os_tick_irqack (void) {
   /* ... */
}
```

Whichever timer you use you must next set up its input clock value. Next, we must define our timer tick rate. This is the rate at which timer interrupts are generated. On each timer tick, the RTOS kernel will run the scheduler to determine if it is necessary to perform a context switch and replace the running thread. The timer tick value will depend on your application, but the default starting value is set to 10 ms.

Timeslice Configuration

The final configuration setting allows you to enable round-robin scheduling and define the timeslice period. This is a multiple of the timer tick rate, so in the above example, each thread will run for five ticks or 50 ms before it will pass execution to another thread of the same priority that is ready to run. If no thread of the same priority is ready to run, it will continue execution. The system configuration options also allow you to enable and configure the virtual timer thread. If you are going to use the virtual timers, this option must be configured or the timers will not work. Then lastly, if you are going to trigger a thread from an interrupt routine using event flags, then it is possible to define a FIFO queue for triggered signals. This buffer signal triggers in the event of bursts of interrupt activity.

Scheduling Options

The CMSIS RTOS allows you to build an application with three different kernel scheduling options. These are round-robin scheduling, preemptive scheduling, and cooperative multitasking. A summary of these options follows.

Preemptive Scheduling

If the round-robin option is disabled in the RTX_Config_CM.c file, each thread must be declared with a different priority. When the RTOS is started and the threads are created, the thread with the highest priority will run.

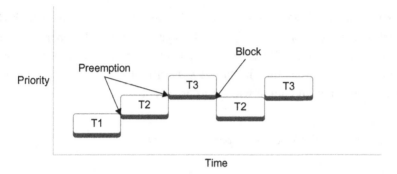

Figure 6.15
In a preemptive RTOS, each thread has a different priority level and will run until it is preempted or has reached a blocking OS call.

This thread will run until it blocks, that is, it is forced to wait for an event flag, semaphore, or other object. When it blocks, the next ready thread with the highest priority will be scheduled and will run until it blocks, or a higher priority thread becomes ready to run. So, with preemptive scheduling, we build a hierarchy of thread execution, with each thread consuming variable amounts of runtime.

Round-Robin Scheduling

A round-robin-based scheduling scheme can be created by enabling the round-robin option in the RTL_Config.c file and declaring each thread with the same priority.

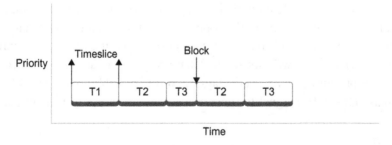

Figure 6.16
In a round-robin RTOS, threads will run for a fixed period or timeslice or until they reach a blocking OS call.

In this scheme, each thread will be allotted a fixed amount of runtime before execution is passed to the next ready thread. If a thread blocks before its timeslice has expired, execution will be passed to the next ready thread.

Round-Robin Preemptive Scheduling

As discussed at the beginning of this chapter, the default scheduling option for the Keil RTX RTOS is round-robin preemptive. For most applications this is the most useful option and you should use this scheduling scheme unless there is a strong reason to do otherwise.

Cooperative Multitasking

A final scheduling option is cooperative multitasking. In this scheme, round-robin scheduling is disabled and each thread has the same priority. This means that the first thread to run will run forever unless it blocks. Then execution will pass to the next ready thread.

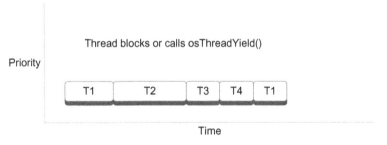

Figure 6.17
In a cooperative RTOS, each thread will run until it reaches a blocking OS call or uses the osThreadYield() function.

Threads can block on any of the standard OS objects, but there is also an additional OS call, osThreadYield, that schedules a thread to the ready state and passes execution to the next ready thread.

Priority Inversion

Finally, no discussion of RTOS scheduling would be complete without mentioning priority inversion.

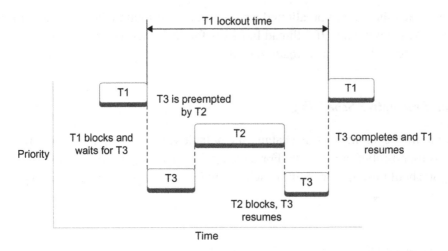

Figure 6.18

A priority inversion is a common RTOS design error. Here a high-priority thread may become delayed or permanently blocked by a medium-priority thread.

In a preemptive scheduling system, it is possible for a high-priority thread T1 to block while it calls a low-priority thread T3 to perform a critical function before T1 continues. However, the low-priority thread T3 could be preempted by a medium priority thread T2. Now T2 is free to run until it blocks (assuming it does) before allowing T3 to resume completing its operation and allowing T1 to resume execution. The upshot is the high-priority thread T1 is blocked and becomes dependent on T2 to complete before it can resume execution.

```
osThreadSetPriority(t_phaseD, osPriorityHigh);  //raise the priority of thread phaseD
  osSignalSet(t_phaseD,0x001);  //trigger it to run//  Call task four to write to the LCD
  osSignalWait(0x01,osWaitForever);  //wait for thread phase D to complete
osThreadSetPriority(t_phaseD,osPriorityBelowNormal);  //lower its priority
```

The answer to this problem is priority elevation. Before T1 calls T3, it must raise the priority of T3 to its level. Once T3 has completed its priority can be lowered back to its initial state.

Exercise: Priority Inversion

In this exercise we will create four threads: two are running at normal priority, one is running at high priority, and one is running at below normal priority. A priority inversion is caused by the high-priority thread waiting for a signal from the below normal task. The normal priority tasks toggle GPIOB port pin 2, while the high and below normal threads toggle GPIOB port pin 1.

Open the project in c:\exercises\CMSIS RTOS priority inversion.

Build the code and start the debugger.

Open the Peripherals\General Purpose IO\GPIOB window.

Run the code.

The main(), phaseA(), and phaseB() tasks are running at normal priority. PhaseA() is waiting for a signal from phaseD(), but phaseD() cannot run because it is at a low priority and will always be preempted by main(), phaseA(), and phaseB().

ID	Name	Priority	State
255	os_idle_demon	0	Ready
5	phaseA	6	Wait_AND
4	phaseD	3	Ready
3	phaseC	4	Wait_AND
2	phaseB	4	Wait_DLY
1	main	4	Running

Examine the code and understand the problem.

Exit the debugger.

Uncomment the priority control code in the phaseA thread.

```
osThreadSetPriority(t_phaseD, osPriorityHigh);
```

Build the project and start the debugger.

Set a breakpoint in the phaseD thread.

```
70 □ void phaseD (void const *argument) {
71 □   for (;;) {
72        osSignalWait(0x01,osWaitForever);
73        Delay(100);
74        LED_Off(3);
75        osSignalSet(t_phaseA,0x01);
76  ├   }
77 }
```

Run the code.

Examine the state of the threads when the code halts.

ID	Name	Priority	State
255	os_idle_demon	0	Ready
5	phaseA	6	Wait_AND
4	phaseD	6	Running
3	phaseC	4	Ready
2	phaseB	4	Ready
1	main	4	Ready

The phaseD thread now has the same priority level as phaseA and a higher priority than phaseB or phaseC.

Remove the breakpoint and run the code.

Both GPIO pins will now toggle as expected.

CHAPTER 7

Practical DSP for the Cortex-M4

This chapter looks at the DSP capabilities of the Cortex-M4. How best to use the DSP intrinsic functions for custom algorithms. ARM also publishes a free DSP library, and this chapter will look at implementing an FFT as well as Infinite Impulse Response (IIR) and Finite Impulse Response (FIR) filters. Techniques for creating real-time DSP applications within an RTOS framework are also presented. The Cortex-M4 may also be fitted with a single precision FPU; it is important to understand how it has been implemented alongside the main Cortex CPU.

Introduction

The Cortex-M4 is the most powerful member of the Cortex-M processor family. While the Cortex-M4 has the same features and performance level of 1.25 DMIPS/MHz as the Cortex-M3, it has a number of architectural enhancements that help dramatically boost its math ability. The key enhancements over the Cortex-M3 are the addition of "single instruction multiple data" or SIMD instructions, an improved MAC unit for integer math, and the optional addition of a single precession FPU. These enhancements give the Cortex-M4 the ability to run DSP algorithms at high enough levels of performance to compete with dedicated 16-bit DSP processors.

Figure 7.1
The Cortex-M4 extends the Cortex-M3 with the addition of DSP instructions and fast math capabilities. This creates a microcontroller capable of supporting real-time DSP algorithms and DSC.

Cortex-M4 Hardware Floating Point Unit

One of the major features of the Cortex-M4 processor is the hardware FPU. The FPU supports single precision floating point arithmetic operations to the IEEE 7xx standard.

Initially, the FPU can be thought of as a coprocessor that is accessed by dedicated instructions to perform most floating point arithmetic operations in a few cycles.

Table 7.1: Floating Point Unit Performance

Operation	Cycle Count
Add/subtract	1
Divide	14
Multiply	1
MAC	3
Fused MAC	3
Square root	14

The FPU consists of a group of control and status registers and 31 single precision scalar registers. The scalar registers can also be viewed as 16 double word registers.

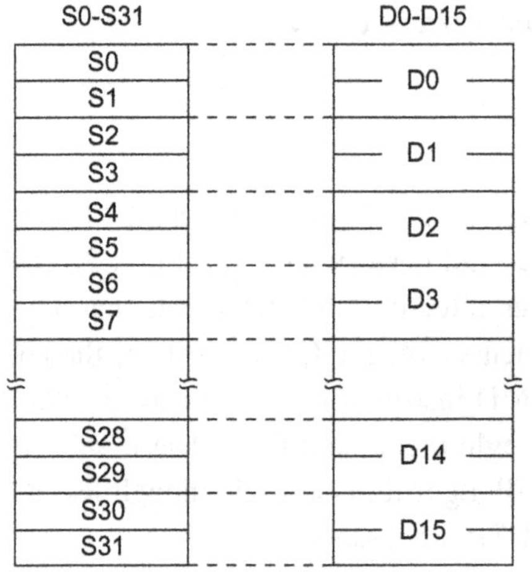

Figure 7.2
The FPU 32-bit scalar registers may also be viewed as 64-bit double word registers. This supports very efficient casting between C types.

While the FPU is designed for floating point operations, it is possible to load and store fixed point and integer values. It is also possible to convert between floating point and fixed point and integer values. This means C casting between floating point and integer values can be done in a cycle.

FPU Integration

While it is possible to consider the FPU as a coprocessor adjacent to the Cortex-M4 processor, this is not really true. The FPU is an integral part of the Cortex-M4 processor; the floating point instructions are executed within the FPU in a parallel pipeline to the Cortex-M4 processor instructions. While this increases the FPU performance, it is "invisible" to the application code and does not introduce any strange side effects.

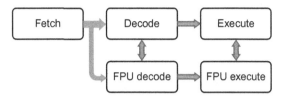

Figure 7.3
The FPU is described as a coprocessor in the register documentation. In reality, it is very tightly coupled to the main instruction pipeline.

FPU Registers

In addition to the scalar registers, the FPU has a block of control and status registers.

Table 7.2: FPU Control Registers

Register	Description
Coprocessor access control	Controls the privilege access level to the FPU
Floating point context control	Configures stacking and lazy stacking options
Floating point context address	Holds the address of the unpopulated FPU stack space
Floating point default status control	Holds the FPU condition codes and FPU configuration options
Floating point status control	Holds the default status control values

All of the FPU registers are memory mapped except the floating point status control register (FPSCR), which is a CPU register accessed by the MRS and MSR instructions. Access to this function is supported by a CMSIS core function:

```
uint32_t __get_FPSCR(void);
void __set_FPSCR (uint32_t fpscr).
```

The FPSCR register contains three groups of bits. The top 4 bits contain condition code flags N, Z, C, V that match the condition code flags in the xPSR. These flags are set and cleared in a similar manner by results of floating point operations. The next groups of bits contain configuration options for the FPU. These bits allow you to change the operation of the FPU from the IEEE 754 standard. Unless you have a strong reason to do this, it is recommended to leave them alone. The final group of bits are status flags for the FPU exceptions. If the FPU encounters an error during execution, an exception will be raised and the matching status flag will be set. The exception line is permanently enabled in the FPU and just needs to be enabled in the NVIC to become active. When the exception is raised, you will need to interrogate these flags to work out the cause of the error. Before returning from the FPU exception, the status flags must be cleared. How this is done depends on the FPU exception stacking method.

Enabling the FPU

When the Cortex-M4 leaves the reset vector, the FPU is disabled. The FPU is enabled by setting the coprocessor 10 and 11 bits in the Co processor Access Control Register (CPARC). It is necessary to use the data barrier instruction to ensure that the write is made before the code continues. The instruction barrier command is also used to ensure that the pipeline is flushed before the code continues.

```
SCB->CPACR |= ((3UL << 10*2)|(3UL << 11*2));  // Set CP10 & CP11 Full Access
__DSB();  //Data barrier
__ISB();  //Instruction barrier
```

In order to write to the CPARC register, the processor must be in privileged mode. Once enabled, the FPU may be used in privileged and unprivileged modes.

Exceptions and the FPU

When the FPU is enabled, an extended stack frame will be pushed and an exception is raised. In addition to the standard stack frame, the Cortex-M4 also pushes the first 16 FPU scalar registers and the FPSCR. This extends the stack frame from 32 to 100 bytes. Clearly pushing this amount of data onto the stack, every interrupt would increase the interrupt latency significantly. To keep the 12 cycle interrupt latency, the Cortex-M4 uses a technique called lazy stacking. When an interrupt is raised, the normal stack frame is pushed onto the stack and the stack pointer is incremented to leave space for the FPU registers, but their values are not pushed onto the stack. This leaves a void space in the stack. The start address of this void space is automatically stored in the floating point context address register (FPCAR). If the interrupt routine uses floating point calculations, the FPU registers will be pushed into this space using the address stored in the FPCAR as a base address. The floating point context control register controls the stacking method used. Lazy stacking is enabled by default when the FPU is first enabled. The stacking method is controlled by the most significant 2 bits in the floating point context control register; these are the Automatic State Preservation Enable (ASPEN) and Lazy State Preservation Enable (LSPEN) bits.

Table 7.3: Lazy Stacking Options

LSPEN	ASPEN	Configuration
0	0	No automatic state preservation; only use when the interrupts do not use floating point
0	1	Lazy stacking disabled
1	0	Lazy stacking enabled
1	1	Invalid configuration

Using the FPU

Once you have enabled the FPU, the compiler will start to use the FPU in place of software libraries. The exception is the square root instruction, which is part of the math.h library. If you have enabled the FPU, the ARM compiler provides an intrinsic instruction to use the FPU square root instruction.

```
float __sqrtf(float x);
```

Note: the intrinsic square root function differs from the ANSI sqrt() library function in that it takes and returns a float rather than a double.

Exercise: Floating Point Unit

This exercise performs a few simple floating point calculations to test out the Cortex-M4 FPU.

Open the project in c:\exercises\FPU.

The code in the main loop is a mixture of math operations to exercise the FPU.

```
#include <math.h>
float a,b,c,d,e;
int f,g = 100;
while(1){
  a = 10.1234;
  b = 100.2222;
  c = a*b;
  d = c-a;
  e = d + b;
  f = (int)a;
  f = f*g;
  a1 = (unsigned int) a;
  a = __sqrtf(e);
  //a = sqrt(e);
```

```
a = c / f;
e = a / 0;
}
}
```

Before we can build this project, we need to make sure that the compiler will build code to use the FPU.

Open the Options for Target window and select the Target menu.

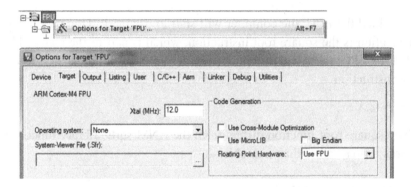

We can enable floating point support by selecting Use FPU in the Floating Point Hardware box. This will enable the necessary compiler options and load the correct simulator model.

Close the Options for Target menu and return to the editor.

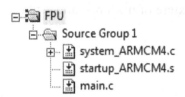

In addition to our source code, the project includes the CMSIS startup and system files for the Cortex-M4.

Now build the project and note the build size.

```
Build target 'FPU'
compiling system_ARMCM4.c...
assembling startup_ARMCM4.s...
compiling main.c...
linking...
Program Size: Code=224 RO-data=224 RW-data=4 ZI-data=1028
"cortexM4.axf" - 0 Error(s), 0 Warning(s).
```

Start the debugger.

When the simulator runs the code to main, it will hit a breakpoint that has been preset in the system_ARMCM4.c file.

```
61   void SystemInit (void)
62 ⊟{
63 ⊟   #if (__FPU_USED == 1)
64         SCB->CPACR |= ((3UL << 10*2) | (3UL << 11*2)  );
65      #endif
66
67      SystemCoreClock = __SYSTEM_CLOCK;
68
69   }
```

The standard microcontroller include file that will define the feature set of the Cortex processor including the availability of the FPU. If the FPU is present on the microcontroller, the SystemInit() function will make sure it is switched on before you reach your application's main function.

```
/* = = = = = = = = = = = = = = = = = = = = = = = = = = = = = = = = = = = = = =*/
/* = = = = = = = = = = Processor and Core Peripheral Section = = = = = = = = = = =*/
/* = = = = = = = = = = = = = = = = = = = = = = = = = = = = = = = = = = = = = =*/

/* ——— Configuration of the Cortex M4 Processor and Core Peripherals ——— */
#define __CM4_REV          0x0001  /*!< Core revision r0p1  */
#define __MPU_PRESENT      1  /*!< MPU present or not  */
#define __NVIC_PRIO_BITS   3  /*!< Number of Bits used for Priority Levels  */
#define __Vendor_SysTickConfig  0  /*!< Set to 1 if different SysTick Config is used  */
#define __FPU_PRESENT      1  /*!< FPU present or not  */
```

Open the main.c module and run the code to the main while() loop.

```
 4   //#include <math.h>
 5 ⊟int main (void){
 6    float a,b,c,d,e;
 7    int f;
 8 ⊟   while(1){
 9       a= 10.123
10       b=100.2222
11       c= a*b;
12       d= c-a;
13       e= d+b;
14       f =(int)a;
15       a=__sqrtf
16       //a= sqrt
```

Split Window horizontally

Go to Headerfile

Show Disassembly at 0x00000120

Set Program Counter

{} Run to Cursor line   Ctrl+F10

Now open the disassembly window.

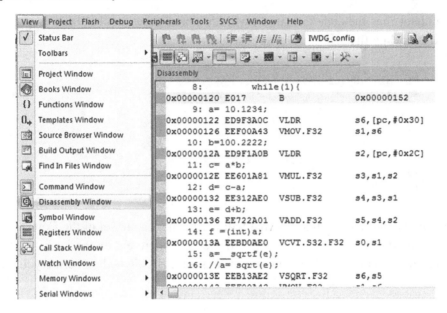

This will show the C source code interleaved with the Cortex-M4 assembly instructions.

In the project window select the registers window.

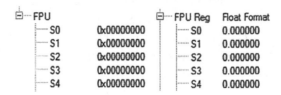

The register window now shows the 31 scalar registers in their raw format and the IEEE 7xx format.

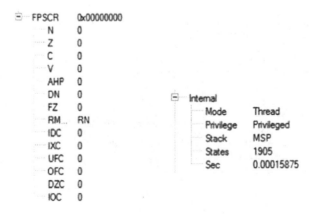

The contents of the FPSCR are also shown. In this exercise, we will also be using the states (cycle count) value also shown in the register window.

Highlight the assembly window and step through each operation noting the cycle time for each calculation.

Now quit the debugger and change the code to use software floating point libraries.

```
a1 = (unsigned int) a;
//a = __sqrtf(e);
a = sqrt(e);
```

Here we need to include the math.h library, comment out the __sqrtf() intrinsic and replace it with the ANSI C sqrt() function.

In the Options for Target\Target settings, we also need to remove the FPU support.

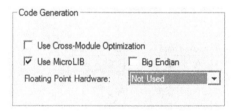

Rebuild the code and compare the build size to the original version.

```
Build target 'FPU'
compiling main.c...
linking...
Program Size: Code=1440 RO-data=224 RW-data=8 ZI-data=1024
"cortexM4.axf" - 0 Error(s), 0 Warning(s).
```

Now restart the debugger run and run the code to main.

In the disassembly window step through the code and compare the number of cycles used for each operation to the number of cycles used by the FPU.

By the end of this exercise, you can clearly see not only the vast performance improvement provided by the FPU but also its impact on project code size. The only downside is the additional cost to use a microcontroller fitted with the FPU and the additional power consumption when it is running.

Cortex-M4 DSP and SIMD Instructions

The Thumb-2 instruction set has a number of instructions that are useful in DSP algorithms.

Table 7.4: Thumb-2 DSP Instructions

Instruction	Description
CLZ	Count leading zeros
REV, REV16, REVSH, and RBIT	Reverse instructions
BFI	Bit field insert
BFC	Bit field clear
UDIV and SDIV	Hardware divide
SXT and UXT	Sign and zero extend

The Cortex-M4 instruction set includes a new group of instructions that can perform multiple arithmetic calculations in a single cycle. The SIMD instructions allow multiple parallel arithmetic operations on 8 or 16 bit data quantites. The 8 and 16 bit values must be packet into two 32 bit words. So, for example, you can perform two 16-bit multiplies and a 32- or 64-bit accumulate or a quad 8-bit addition in one processor cycle. Since many DSP algorithms work on a pipeline of data, the SIMD instructions can be used to dramatically boost performance.

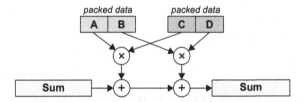

Figure 7.4
The SIMD instructions support multiple arithmetic operations in a single cycle. The operand data must be packed into 32-bit words.

The SIMD instructions have an additional field in the xPSR register. The "greater than or equal" (GE) field contains 4 bits, which correspond to the 4 bytes in the SIMD instruction result operand. If the result operand byte is GE to zero then the matching GE flag will be set.

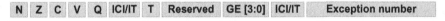

Figure 7.5
The Cortex-M4 xPSR register has an additional GE field. Each of the four GE bits are updated when an SIMD instruction is executed.

The SIMD instructions can be considered as three distinct groups: add and subtract operations, multiply operations, and supporting instructions. The add and subtract operations can be performed on 8- or 16-bit signed and unsigned quantities. A signed and unsigned halving instruction is also provided; this instruction adds or subtracts the 8- or 16-bit quantities and then halves the result as shown in the following.

Table 7.5: SIMD Add Halving and Subtract Halving Instructions

Instruction	Description	Operation
UHSUB16	Unsigned halving 16-bit subtract	Res[15:0] = (Op1[15:0] − Op2[15:0])/2 Res[31:16] = (Op1[31:16] − Op2[31:16])/2
UHADD16	Unsigned halving 16-bit add	Res[15:0] = (Op1[15:0] + Op2[15:0])/2 Res[31:16] = (Op1[31:16] + Op2[31:16]/2

The SIMD instructions also include an add and subtract with exchange (ASX) and a subtract and add with exchange (SAX). These instructions perform an add and subtract on the two halfwords and store the results in the upper and lower halfwords of the destination register.

Table 7.6: SIMD Add Exchange and Subtract Exchange Instructions

Instruction	Description	Operation
USAX	Unsigned 16-bit subtract and add with exchange	Res[15:0] = Op1[15:0] + Op2[31:16] Res[31:16] = Op1[31:16] − Op2[15:0]
UASX	Unsigned 16-bit add and subtract with exchange	Res[15:0] = Op1[15:0] + Op2[31:16] Res[31:16] = Op1[31:16] − Op2[15:0]

A further group of instructions combine these two operations in a subtract and add (or add and subtract) with exchange halving instruction. This gives quite few possible permutations. A summary of the add and subtract SIMD instructions is shown below.

Table 7.7: Permutations of the SIMD Add, Subtract, Halving, and Saturating Instructions

Prefix / Instruction	S Signed	Q Signed Saturating	SH Signed Halving	U Unsigned	UQ Unsigned Saturating	UH Unsigned Halving
ADD8	SADD8	QADD8	SHADD8	UADD8	UQADD8	UHADD8
SUB8	SSUB8	QSUB8	SHSUB8	USUB8	UQSUB8	UHSUB8
ADD16	SADD16	QADD16	SHADD16	UADD16	UQADD16	UHADD16
SUB16	SSUB16	QSUB16	SHSUB16	USUB16	UQSUB16	UHSUB16
ASX	SASX	QASX	SHASX	UASX	UQASX	UHASX
SAX	SSAX	QSAX	SHSAX	USAX	UQSAX	UHSAX

The SIMD instructions also include a group of multiply instructions that operate on packed 16-bit signed values. Like the add and subtract instructions, the multiply instructions also support saturated values. As well as multiply and multiply accumulate the SIMD multiply instructions support multiply subtract and multiply add as shown in the following.

Table 7.8: SIMD Multiply Instructions

Instruction	Description	Operation
SMLAD	Q setting dual 16-bit signed multiply with single 32-bit accumulator	$X = X + (A \times B) + (C \times D)$
SMLALD	Dual 16-bit signed multiply with single 64-bit accumulator	$X = X + (A \times B) + (C \times D)$
SMLSD	Q setting dual 16-bit signed multiply subtract with 32-bit accumulator	$X = X + (A \times B) - (B \times C)$
SMLSLD	Q setting dual 16-bit signed multiply subtract with 64-bit accumulator	$X = X + (A \times B) - (B \times C)$
SMUAD	Q setting sum of dual 16-bit signed multiply	$X = (A \times B) + (C \times D)$
SMUSD	Dual 16-bit signed multiply returning difference	$X = (A \times B) - (C \times D)$

To make the SIMD instructions more efficient, a group of supporting pack and unpack instructions have also been added. The pack\unpack instructions can be used to extract 8- and 16-bit values from a register and move them to a destination register. The unused bits in the 32-bit word can be set to zero (unsigned) or one (signed). The pack instructions can also take two 16-bit quantities and load them into the upper and lower halfwords of a destination register.

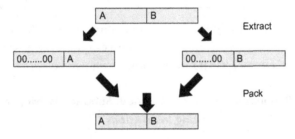

Figure 7.6
The SIMD instruction group includes support instructions to pack 32-bit words with 8- and 16-bit quantities.

Table 7.9: SIMD Supporting Instructions

Mnemonic	Description
PKH	Pack halfword
SXTAB	Extend 8-bit signed value to 32 bits and add
SXTAB16	Dual extend 8-bit signed value to 16 bits and add
SXTAH	Extend 16-bit signed value to 32 bits and add
SXTB	Sign extend a byte
SXTB16	Dual extend 8-bit signed value to 16 bits and add
SXTH	Sign extend a halfword
UXTAB	Extend 8-bit signed value to 32 bits and add
UXTAB16	Dual extend 8 to 16 bits and add
UXTAH	Extend a 16-bit value and add
UXTB	Zero extend a byte
UXTB16	Dual zero extend 8 to 16 bits and add
UXTH	Zero extend a halfword

When an SIMD instruction is executed, it will set or clear the xPSR GE bits depending on the values in the resulting bytes or halfwords. An additional select (SEL) instruction is provided to access these bits. The SEL instruction is used to select bytes or halfwords from two input operands depending on the condition of the GE flags.

Table 7.10: xPSR "GE" Bit Field Results

GE Bit[3:0]	GE Bit = 1	GE Bit = 0
0	Res[7:0] = OP1[7:0]	Res[7:0] = OP2[7:0]
1	Res[15:8] = OP1[15:8]	Res[15:8] = OP2[15:8]
2	Res[23:16] = OP1[23:16]	Res[23:16] = OP2[23:16]
3	Res[31:24] = OP1[31:24]	Res[31:24] = OP2[31:24]

Exercise: SIMD Instructions

In this exercise, we will have a first look at using the Cortex-M4 SIMD instructions. In this exercise, we will simply multiply and accumulate two 16-bit arrays first using an SIMD instruction and then using the standard add instruction.

First open the CMSIS core documentation and the SIMD signed multiply accumulate intrinsic __SMLAD.

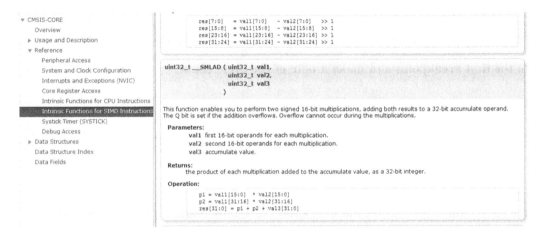

Open the project in c:\exercises\CMSIS core SIMD.

The application code defines two sets of arrays as a union of 16- and 32-bit quantities.

```
union _test{
int16_t  Arry_halfword[100];
int32_t  Arry_word[50];
};
```

The code first initializes the arrays with the values 0–100.

```
for(n = 0;n < 100;n++){
op1.Arry_halfword[n] = op2.Arry_halfword[n] = n;  }
```

Then multiply accumulates first using the SIMD instruction, then the standard multiply accumulate.

```
for(n = 0;n < 50;n++){
Result = __SMLAD(op1.Arry_word[n],op2.Arry_word[n],Result);}
```

The result is then reset and the calculation is repeated without using the SIMD instruction.

```
Result = 0;
for(n = 0;n < 100;n++){
Result = Result + (op1.Arry_halfword[n] * op2.Arry_halfword[n]);}
```

Build the code and start the debugger.

Set a breakpoint at lines 23 and 28.

```
● 23 ⊟   for (n=0;n<50;n++) {
  24
  25      Result = __SMLAD(op1.Arry_word[n],op2.Arry_word[n],Result);
  26
  27 ⊦ }
● 28 │ Result = 0;
```

Run to the first breakpoint and make a note of the cycle count.

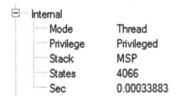

```
⊟ Internal
    Mode        Thread
    Privilege   Privileged
    Stack       MSP
    States      4066
    Sec         0.00033883
```

Run the code until it hits the second breakpoint. See how many cycles have been used to execute the SIMD instruction.

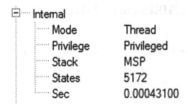

```
⊟ Internal
    Mode        Thread
    Privilege   Privileged
    Stack       MSP
    States      5172
    Sec         0.00043100
```

Cycles used = 5172−4066 = 1106

Set a breakpoint at the final while loop.

```
30 □for(n=0;n<100;n++){
31     Result = Result + (op1.Arry_halfword[n] * op2.Arry_halfword[n]);
32   }
33 □while(1){
```

Run the code and see how many cycles are used to perform the calculation without using the SIMD instruction.

Internal	
Mode	Thread
Privilege	Privileged
Stack	MSP
States	7483
Sec	0.00062358

Cycles used $= 7483 - 5172 = 2311$

Compare the number of cycles used to perform the same calculation without using the SIMD instructions.

As expected, the SIMD instructions are much more efficient when performing calculations on large data sets.

The primary use for the SIMD instructions is to optimize the performance of DSP algorithms. In the next exercise, we will look at various techniques in addition to the SIMD instructions that can be used to boost the efficiency of a given algorithm.

Exercise: Optimizing DSP Algorithms

In this exercise, we will look at optimizing an FIR filter. This is a classic algorithm that is widely used in DSP applications.

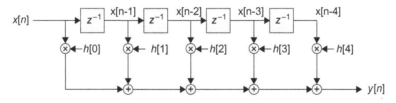

Figure 7.7
An FIR filter is an averaging filter with its characteristics defined by a series of coefficients applied to each sample in a series of "taps."

The FIR filter is an averaging filter that consists of a number of "taps." Each tap has a coefficient and as the filter runs each sample is multiplied against the coefficient in the first tap and then shifted to the next tap to be multiplied against its coefficient when the next sample arrives. The output of each tap is summed to give the filter output. Or to put it mathematically,

$$y[n] = \sum_{k=0}^{N-1} h[k]x[n-k]$$

Open the project in c:\exercises\FIR_optimization.

Build the project and start the debugger.

The main function consists of four FIR functions that introduce different optimizations to the standard FIR algorithm.

```
int main (void){
fir (data_in,data_out, coeff,&index, FILTERLEN, BLOCKSIZE);
fir_block (data_in,data_out, coeff,&index, FILTERLEN, BLOCKSIZE);
fir_unrolling (data_in,data_out, coeff,&index, FILTERLEN, BLOCKSIZE);
fir_SIMD (data_in,data_out, coeff,&index, FILTERLEN, BLOCKSIZE);
fir_SuperUnrolling (data_in,data_out, coeff,&index, FILTERLEN, BLOCKSIZE);
while(1);
}
```

Step into the first function and examine the code.

The filter function is implemented in C as shown below. This is a standard implementation of an FIR filter written purely in C.

```
void fir(q31_t *in, q31_t *out, q31_t *coeffs, int *stateIndexPtr,
            int filtLen, int blockSize)
{
  int sample;
  int k;
  q31_t sum;
  int stateIndex = *stateIndexPtr;
  for(sample = 0; sample<blockSize; sample++)
    {
      state[stateIndex++] = in[sample];
      sum = 0;
      for(k = 0;k<filtLen;k++)
        {
```

```
            sum + = coeffs[k] * state[stateIndex];
            stateIndex-;
            if (stateIndex<0)
               {
                 stateIndex = filtLen-1;
               }
          }
       out[sample] = sum;
     }
     *stateIndexPtr = stateIndex;
   }
```

While this compiles and runs fine, it does not take full advantage of the Cortex-M4 DSP enhancements. To get the best out of the Cortex-M4, we need to optimize this algorithm, particularly the inner loop.

The inner loop performs the FIR multiply and accumulate for each tap.

```
for(k = 0;k<filtLen;k++)
  {
    sum + = coeffs[k] * state[stateIndex];
    stateIndex-;
    if (stateIndex<0)
       {
         stateIndex = filtLen-1;
       }
  }
```

The inner loop processes the samples by implementing a circular buffer in the software. While this works fine, we have to perform a test for each loop to wrap the pointer when it reaches the end of the buffer.

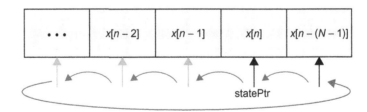

Figure 7.8
Processing data in a circular buffer requires the Cortex-M4 to check for the end of the buffer on each iteration. This increases the execution time.

Run to the start of the inner loop and set a breakpoint.

```
16          for(k=0;k<filtLen;k++)
17      {
18        sum += coeffs[k] * state[stateIndex];
19        stateIndex--;
20        if (stateIndex < 0)
21          {
```

Run the code so it does one iteration of the inner loop and note the number of cycles used.

Circular addressing requires us to perform an end of buffer test on each iteration. A dedicated DSP device can support circular buffers in hardware without any such overhead, so this is one area that we need to improve. By passing our FIR filter function, a block of data rather than individual samples allows us to use block processing as an alternative to circular addressing. This improves the efficiency on the critical inner loop.

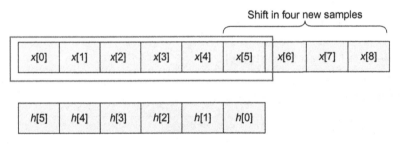

Figure 7.9
Block processing not only increases the size of the buffer but also increases the efficiency of the inner processing loop.

By increasing the size of the state buffer to number of filter taps + processing block size, we can eliminate the need for circular addressing. In the outer loop, the block of samples is loaded into the top of the state buffer.

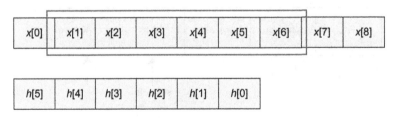

Figure 7.10
With block processing, the fixed size buffer is processed without the need to check for the end of the buffer.

The inner loop then performs the filter calculations for each sample of the block by sliding the filter window one element to the right for each pass through the loop. So the inner loop now becomes

```
for(k = 0; k < filtLen; k++)
{
    sum + = coeffs[k] * state[stateIndex];
    stateIndex++;
}
```

Once the inner loop has finished processing, the current block of sample data held in the state buffer must be shifted to the right and a new block of data should be loaded.

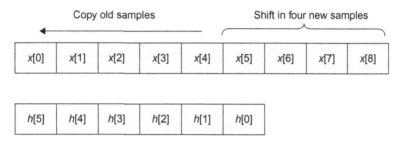

Figure 7.11
Once the block of data has been processed, the outer loop shifts the samples one block to the left and adds a new block of data.

Now step into the second FIR function.

Examine how the code has been modified to process blocks of data.

There is some extra code in the outer loop, but this is only executed once per tap and becomes insignificant compared to the savings made within the inner loop particularly for large block sizes.

Set a breakpoint on the same inner loop and record the number of cycles it takes to run.

Next we can further improve the efficiency of the inner loop by using a compiler trick called "loop unrolling." Rather than iterating round the loop for each tap, we can process several taps in each iteration by inlining multiple tap calculations per loop.

```
I = filtLen >> 2
for(k = 0;k < filtLen;k++)
{
    sum + = coeffs[k] * state[stateIndex];
    stateIndex++;
```

```
    sum + = coeffs[k] * state[stateIndex];
    stateIndex++;
    sum + = coeffs[k] * state[stateIndex];
    stateIndex++;
    sum + = coeffs[k] * state[stateIndex];
    stateIndex++;
  }
```

Now step into the third FIR function.

Set a breakpoint on the same inner loop and record the number of cycles it takes to run. Divide this by four and compare it to the previous implementations.

The next step is to make use of the SIMD instructions. By packing the coefficient and sample data into 32-bit words, the single signed multiply accumulates can be replaced by dual signed multiply accumulates, which allows us to extend the loop unrolling from four summations to eight for the same number of cycles.

```
  for(k = 0;k < filtLen;k++)
  {
    sum + = coeffs[k] * state[stateIndex];
    stateIndex++;
    sum + = coeffs[k] * state[stateIndex];
    stateIndex++;
    sum + = coeffs[k] * state[stateIndex];
    stateIndex++;
    sum + = coeffs[k] * state[stateIndex];
    stateIndex++;
  }
```

Step into the fourth FIR function and again calculate the number of cycles used per tap for the inner loop.

Remember we are now calculating eight summations, so divide the raw loop cycle count by eight.

To reduce the cycle count of the inner loop even further, we can extend the loop unrolling to calculate several results simultaneously.

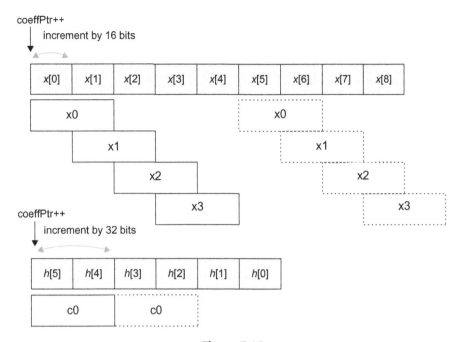

Figure 7.12
Super loop unrolling extends loop unrolling to process multiple output samples simultaneously.

This is kind of a "super loop unrolling" where we perform each of the inner loop calculations for a block of data in one pass.

```
sample = blockSize/4;
  do
  {
    sum0 = sum1 = sum2 = sum3 = 0;
    statePtr = stateBasePtr;
    coeffPtr = (q31_t *)(S->coeffs);
    x0 = *(q31_t *)(statePtr++);
    x1 = *(q31_t *)(statePtr++);
    i = numTaps >> 2;
    do
    {
      c0 = *(coeffPtr++);
      x2 = *(q31_t *)(statePtr++);
      x3 = *(q31_t *)(statePtr++);
      sum0 = __SMLALD(x0, c0, sum0);
      sum1 = __SMLALD(x1, c0, sum1);
      sum2 = __SMLALD(x2, c0, sum2);
```

```
        sum3 = __SMLALD(x3, c0, sum3);
        c0 = *(coeffPtr++);
        x0 = *(q31_t *)(statePtr++);
        x1 = *(q31_t *)(statePtr++);
        sum0 = __SMLALD(x0, c0, sum0);
        sum1 = __SMLALD(x1, c0, sum1);
        sum2 = __SMLALD (x2, c0, sum2);
        sum3 = __SMLALD (x3, c0, sum3);
    } while(-i);
    *pDst++ = (q15_t) (sum0 >> 15);
    *pDst++ = (q15_t) (sum1 >> 15);
    *pDst++ = (q15_t) (sum2 >> 15);
    *pDst++ = (q15_t) (sum3 >> 15);
    stateBasePtr = stateBasePtr + 4;
} while(-sample);
```

Now step into the final FIR function and again calculate the number of cycles used by the inner loop per tap.

This time we are calculating eight summations for four taps simultaneously; this brings us close to one cycle per tap, which is comparable to a dedicated DSP device.

While you can code DSP algorithms in C and get reasonable performance, these kinds of optimizations are needed to get performance levels comparable to a dedicated DSP device. This kind of code development requires experience with the Cortex-M4 and the DSP algorithms. Fortunately, ARM provides a free DSP library already optimized for the Cortex-M4.

The CMSIS DSP Library

While it is possible to code all of your own DSP functions, this can be time consuming and requires a lot of domain specific knowledge. To make it easier to add common DSP functions to your application, ARM have published a library of 61 common DSP functions as part of CMSIS. Each of these functions are optimized for the Cortex-M4 but can also be compiled to run on the Cortex-M3 and even on the Cortex-M0. The CMSIS DSP library is a free download and is licensed for use in any commercial or noncommercial project. ·
The CMSIS DSP library is also included as part of the MDK-ARM installation and just needs to be added to your project. The installation includes a prebuilt library for each of the Cortex-M processors and all the source code.

Documentation for the library is included as part of the CMSIS help, which can be found in the Books tab of the μVision project window.

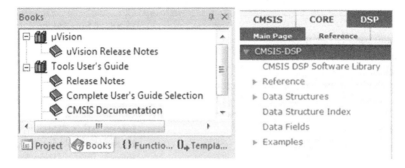

Figure 7.13
The CMSIS DSP documentation is available through the µVision Books tab.

CMSIS DSP Library Functions

The Cortex-M DSP library provides easy to use functions for the most commonly used DSP algorithms. The functions included in the library are shown below.

Table 7.11: CMSIS DSP Library Functions

Basic math functions	Matrix functions
Vector multiplication	Matrix initialization
Vector subtraction	Matrix addition
Vector addition	Matrix subtraction
Vector scale	Matrix multiplication
Vector shift	Matrix inverse
Vector offset	Matrix transpose
Vector negate	Matrix scale
Vector absolute	**Transforms**
Vector dot product	Complex FFT functions
Fast math functions	Real FFT functions
Cosine	DCT type IV functions
Sine	**Controller functions**
Square root of number	**Sine cosine**
Complex math functions	**PID**
Complex conjugate	Vector park transform
Complex dot product	Vector inverse park transform
Complex magnitude	Vector Clarke transform
Complex magnitude squared	Vector inverse Clarke transform
Complex by complex multiplication	**Statistical functions**
Complex by real multiplication	Power
Filters	Root mean square
Convolution	Standard deviation
Partial convolution	Variance
Correlation	Maximum
FIR filter	Minimum
FIR decemation	Mean

(Continued)

Table 7.11: (Continued)

FIR lattice filter	**Support functions**
Infinite impulse response lattice filter	Vector copy
FIR sparse filter	Vector fill
FIR filter interpolation	Convert 8-bit integer value
Biquad cascade IIR filter using direct form I structure	Convert 16-bit integer value
Biquad cascade IIR filter 32 × 64 using direct form I structure	Convert 32-bit integer value
Biquad cascade IIR filter using direct form II transposed structure	Convert 32-bit FPU
Least mean squares FIR filter	**Interpolation functions**
Least mean squares normalized FIR filter	Linear interpolation function
	Bilinear interpolation function

Exercise: Using the Library

In this project, we will have a first look at using the CMSIS DSP library by setting up a project to experiment with the PID control algorithm.

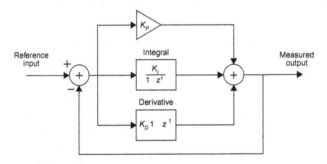

Figure 7.14
A PID control loop consists of proportional, integral, and derivative control blocks.

Open the project in c:\exercises\PID.

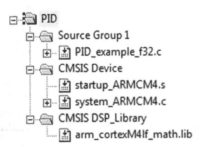

The project is targeted at a Cortex-M4 with FPU. It includes the CMSIS startup and system files for the Cortex-M4 and the CMSIS DSP functions as a precompiled library. The CMSIS DSP library is located in c:\CMSIS\lib with subdirectories for the ARM compiler and GCC versions of the library.

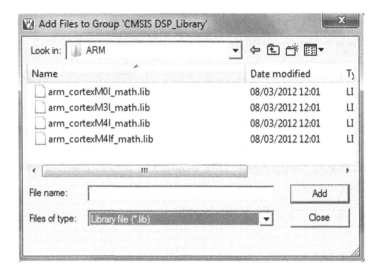

There are precompiled versions of the DSP library for each Cortex processor. There is also a version of the library for the Cortex-M4 with and without the FPU. For our project, the Cortex-M4 floating point library has been added.

Open the Project\Options for Target dialog.

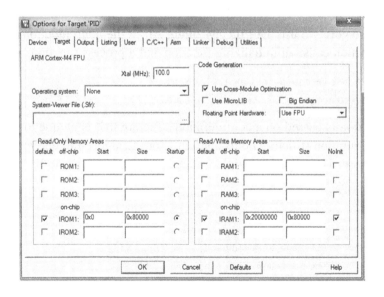

In this example, we are using a simulation model for the Cortex-M4 processor only. A 512 K memory region has been defined for code and data. The FPU is enabled and the CPU clock has been set to 100 MHz.

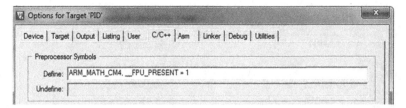

In the compiler options tab, the _FPU_PRESENT define is set to one. Normally you will not need to configure this option as this will be done in the microcontroller include file. We also need to add a #define to configure the DSP header file for the processor we are using. The library defines are

```
ARM_MATH_CM4
ARM_MATH_CM3
ARM_MATH_CM0
```

The source code for the library is located in c:\keil\arm\CMSIS\DSP_Lib\Source.

Open the file PID_example_f32.c which contains the application code.

To access the library, we need to add its header file to our code.

```
#include "arm_math.h"
```

This header file is located in c:\keil\arm\cmsis\include.

In this project, we are going to use the PID algorithm. All of the main functions in the DSP library have two function calls: an initializing function and a process function.

```
void arm_pid_init_f32 (arm_pid_instance_f32 *S,int32_t resetStateFlag)
__STATIC_INLINE void arm_pid_f32 (arm_pid_instance_f32 *s, float32_t in)
```

The initializing function is passed on a configuration structure that is unique to the algorithm. The configuration structure holds constants for the algorithm, derived values, and arrays for state memory. This allows multiple instances of each function to be created.

```
typedef struct
  {
    float32_t A0;      /**< The derived gain, A0 = Kp + Ki + Kd. */
    float32_t A1;      /**< The derived gain, A1 = -Kp - 2 Kd. */
    float32_t A2;      /**< The derived gain, A2 = Kd. */
    float32_t state[3];    /**< The state array of length 3. */
    float32_t Kp;      /**< The proportional gain. */
    float32_t Ki;      /**< The integral gain. */
    float32_t Kd;      /**< The derivative gain. */
  } arm_pid_instance_f32;
```

The PID configuration structure allows you to define values for the proportional, integral, and derivative gains. The structure also includes variables for the derived gains A0, A1, and A2 as well as a small array to hold the local state variables.

```
int32_t main(void){
int i; S.Kp = 1; S.Ki = 1; S.Kd = 1;
setPoint = 10;
arm_pid_init_f32  (&S,0);
while(1){
  error = setPoint - motorOut;
  motorIn = arm_pid_f32  (&S,error);
  motorOut = transferFunction(motorIn,time);
  time + = 1;
  for(i = 0;i < 100000;i++);
  }}
```

The application code sets the PID gain values and initializes the PID function. The main loop calculates the error value before calling the PID process function. The PID output is fed into a simulated hardware transfer function. The output of the transfer function is fed back into the error calculation to close the feedback loop. The time variable provides a pseudo-time reference.

Build the project and start the debugger.

Add the key variables to the logic analyzer and start running the code.

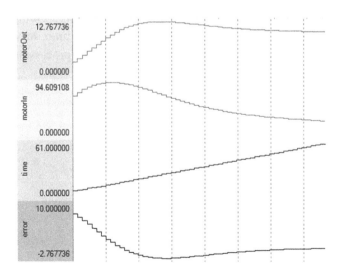

The logic analyzer is invaluable for visualizing data in a real-time algorithm and can be used in the simulator or can capture data from the CoreSight data watch trace unit.

Experiment with the gain values to tune the PID function.

The performance of the PID control algorithm is tuned by adjusting the gain values. As a rough guide, each of the gain values has the following effect.

```
Kp  Effects the rise time of the control signal.
Ki  Effects the steady state error of the control signal.
Kd  Effects the overshoot of the control signal.
```

DSP Data Processing Techniques

One of the major challenges of a DSP application is managing the flow of data from the sensors and ADC through the DSP algorithm and back out to the real world via the DAC.

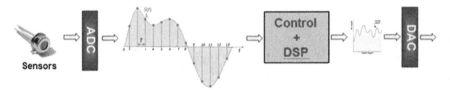

Figure 7.15
A typical DSP system consists of an analog sample stage, microcontroller with DSP algorithm, and an output DAC. In this chapter, we concentrate on the microcontroller software in isolation from the hardware design.

In a typical system, each sampled value is a discrete value at a point in time. The sample rate must be at least twice the signal bandwidth or up to four times the bandwidth for a high-quality oversampled audio system. Clearly the volume of data is going to ramp up very quickly and it becomes a major challenge to process the data in real time. In terms of processing the sampled data, there are two basic approaches, stream processing and block processing.

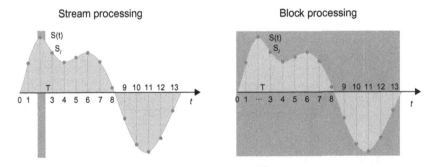

Figure 7.16
Analog data can be processed as single samples with minimum latency or as a block of samples for maximum processing efficiency.

In stream processing, each sampled value is processed individually. This gives the lowest signal latency and also the minimum memory requirements. However, it has the disadvantage of making the DSP algorithm more complex. The DSP algorithm has to be run every time an ADC conversion is made, which can cause problems with other high-priority interrupt routines.

The alternative to stream processing is block processing. Here a number of ADC results are stored in a buffer, typically about 32 samples, and then this buffer is processed by the DSP algorithm as a block of data. This lowers the number of times that the DSP algorithm has to run. As we have seen in the optimization exercise, there are a number of techniques that can improve the efficiency of an algorithm when processing a block of data. Block processing also integrates well with the microcontroller DMA unit and an RTOS. On the downside, block processing introduces more signal latency and requires more memory than stream processing. For the majority of applications, block processing should be the preferred route.

Exercise: FIR Filter with Block Processing

In this exercise, we will implement an FIR filter, this time by using the CMSIS DSP functions. This example uses the same project template as the PID program. The characteristics of the filter are defined by the filter coefficients. It is possible to calculate the coefficient values manually or by using a design tool. Calculating the coefficients is outside the scope of this book, but the appendices list some excellent design tools and DSP books for further reading.

Open the project in c:\examples CMSIS_FIR.

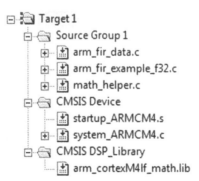

There is an additional data file that holds a sampled data set and an additional math_helper.c file that contains some ancillary functions.

```
int32_t main(void)
{
  uint32_t i;
  arm_fir_instance_f32 S;
  arm_status status;
  float32_t *inputF32, *outputF32;
  /* Initialize input and output buffer pointers */
  inputF32 = &testInput_f32_1 kHz_15 kHz[0];
  outputF32 = &testOutput[0];
  /* Call FIR init function to initialize the instance structure. */
  arm_fir_init_f32(&S, NUM_TAPS, (float32_t *)&firCoeffs32[0], &firStateF32[0],
blockSize);
  for(i = 0; i < numBlocks; i++)
    {
      arm_fir_f32(&S, inputF32 + (i * blockSize), outputF32 + (i * blockSize), blockSize);
    }
  snr = arm_snr_f32(&refOutput[0], &testOutput[0], TEST_LENGTH_SAMPLES);
  if (snr < SNR_THRESHOLD_F32)
    {
      status = ARM_MATH_TEST_FAILURE;
    } else {
      status = ARM_MATH_SUCCESS;
    }
```

The code first creates an instance of a 29-tap FIR filter. The processing block size is 32 bytes. An FIR state array is also created to hold the working state values for each tap. The size of this array is calculated as the block size + number of taps − 1. Once the filter has been initialized, we can pass it to the sample data in 32-byte blocks and store the resulting processed data in the output array. The final filtered result is then compared to a precalculated result.

Build the project and start the debugger.

Step through the project to examine the code.

Look up the CMSIS DSP functions used in the help documentation.

Set a breakpoint at the end of the project.

```
209    if( status != ARM_MATH_SUCCESS)
210 ⊟  {
211        while(1);
212    }
213 ├
● 214      while(1);
215 │ }
```

Reset the project and run the code until it hits the breakpoint.

Now look at the cycle count in the registers window.

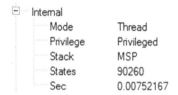

⊟ Internal	
Mode	Thread
Privilege	Privileged
Stack	MSP
States	90260
Sec	0.00752167

Open the project in c:\exercises\CMSIS _FIR\CM3.

This is the same project built for the Cortex-M3.

Build the project and start the debugger.

Set a breakpoint in the same place as the previous Cortex-M4 example.

Reset the project and run the code until it hits the breakpoint.

Now compare the cycle count used by the Cortex-M3 to the Cortex-M4 version.

⊟ Internal	
Mode	Thread
Privilege	Privileged
Stack	MSP
States	855964
Sec	0.07133033

When using floating point numbers, the Cortex-M4 is nearly an order faster than the Cortex-M3 using software floating point libraries. When using fixed point math, the Cortex-M4 still has a considerable advantage over the Cortex-M3.

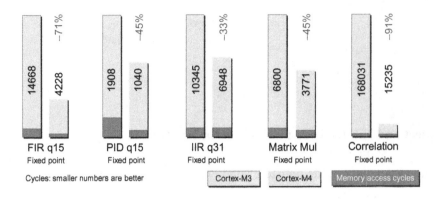

Figure 7.17
The bar graphs show a comparison between the Cortex-M3 and Cortex-M4 for common DSP algorithms.

Fixed Point DSP with Q Numbers

The functions in the DSP library support floating point and fixed point data. The fixed point data is held in a Q number format. Q numbers are fixed point fractional numbers held in integer variables. A Q number has a sign bit followed by a fixed number of bits to represent the integer value. The remaining bits represent the fractional part of the number.

Signed = S IIIIIIIIII.FFFFF

A Q number is defined as the number of bits in the integer portion of the variable and the number of bits in the fractional portion of the variable. Signed Q numbers are stored as two compliment values. A Q number is typically referred to by the number of fractional bits it uses so Q10 has 10 fractional places. The CMSIS DSP library functions are designed to take input values between $+1$ and -1. The supported integer values are shown below.

Table 7.12: Floating Point Unit Performance

CMSIS DSP Typedef	Q Number
Q31_t	Q31
Q15_t	Q15
Q7_t	Q7

The library includes a group of conversion functions to change between floating point numbers and the integer Q numbers. Support functions are also provided to convert between different Q number resolutions.

Table 7.13: CMSIS DSP Type Conversion Functions

arm_float_to_q31	arm_q31_to_float	arm_q15_to_float	arm_q7_to_float
arm_float_to_q15	arm_q31_to_q15	arm_q15_to_q31	arm_q7_to_q31
arm_float_to_q7	arm_q31_to_q7	arm_q15_to_q7	arm_q7_to_q15

As a real-world example, you may be sampling data using a 12-bit ADC that gives an output from 1-0xFFF. In this case, we would need to scale the ADC result to be between $+1$ and -1 and then convert to a Q number.

```
Q31_t ADC_FixedPoint;
float  temp;
```

Read the ADC register to a float variable. Scale between $+1$ and -1.

```
temp = ((float32_t)((ADC_DATA_REGISTER) & 0xFFF) / (0xFFF / 2)) - 1;
```

Convert the float value to a fixed point Q31 value.

```
arm_float_to_q31(&ADC_FixedPoint, &temp, 1);
```

Similarly, after the DSP function has run, it is necessary to convert back to a floating point value before using the result. Here we are converting from a Q31 result to a 10-bit integer value prior to outputting the value to a DAC peripheral.

```
arm_q31_to_float(&temp, &DAC_Float, 1);
DAC_DATA_REGISTER = (((uint32_t)((DAC_Float))) & 0x03FF);
```

Exercise: Fixed Point FFT

In this project, we will use the FFT to analyze a signal.

Open the project in c:\exercises\FFT.

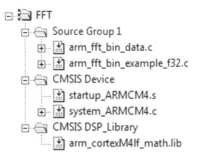

The FFT project uses the same template as the PID example with the addition of a data file that holds an array of sample data. This data is a 10 KHz signal plus random "white noise."

```
int32_t main(void) {
  arm_status status;
  arm_cfft_radix4_instance_q31 S;
  q31_t maxValue;
  status = ARM_MATH_SUCCESS;
  /* Convert the floating point values to the Q31 fixed point format */
  arm_float_to_q31 (testInput_f32_10khz,testInput_q31_10khz,2048);
  /* Initialize the CFFT/CIFFT module */
  status = arm_cfft_radix4_init_q31(&S, fftSize, ifftFlag, doBitReverse);
  /* Process the data through the CFFT/CIFFT module */
  arm_cfft_radix4_q31(&S, testInput_q31_10khz);
  /* Process the data through the Complex Magnitude Module for
  calculating the magnitude at each bin */
  arm_cmplx_mag_q31(testInput_q31_10khz, testOutput, fftSize);
  /* Calculates maxValue and returns corresponding BIN value */
  arm_max_q31(testOutput, fftSize, &maxValue, &testIndex);
  if(testIndex! = refIndex)
  {
    status = ARM_MATH_TEST_FAILURE;
  }
```

The code initializes the complex FFT with a block size of 1024 output bins. When the FFT function is called, the sample data set is first converted to fixed point Q31 format and is then passed to the transform as one block. Next the complex output is converted to a scalar magnitude and then scanned to find the maximum value. Finally, we compare this value to an expected result.

Work through the project code and look up the CMSIS DSP functions used.

Note: In an FFT, the number of output bins will be half the size of the sample data size.

Build the project and start the debugger.

Designing for Real-Time Processing

So far, we have looked at the DSP features of the Cortex-M4 and the supporting DSP library. In this next section, we will look at developing a real-time DSP program that can also support non-real-time features like a user interface or communication stack without disturbing the performance of the DSP algorithm.

Buffering Techniques: The Double or Circular Buffer

When developing your first real-time DSP application, the first decision you have to make is how to buffer the incoming blocks of data from the ADC and the outgoing blocks of data to the DAC. A typical solution is to use a form of double buffer as shown below.

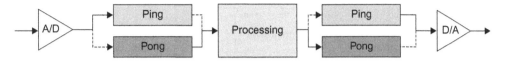

Figure 7.18
A simple DSP system can use a double buffer to capture the stream of data.

While the ping buffer is being filled with data from the ADC, the DSP algorithm will be processing the data stored in the pong buffer. Once the ADC reaches the end of the ping buffer, it will start to store data into the pong buffer and the new data in the ping buffer may be processed. A similar ping-pong structure can be used to buffer the output data to the DAC.

Ping	Pong	Ping	Pong	Ping
Processing	Processing	Processing	Processing	Processing
Ping	Pong	Ping	Pong	Ping

Signal delay = 2 × Block size

Figure 7.19
The double buffer causes minimum signal latency but requires the DSP algorithm to run frequently. This makes it hard to maintain real-time performance when other software routines need to be executed.

This method works well for simple applications; it requires the minimum amount of memory and introduces the minimal latency. The application code must do all of the necessary buffer management. However, as the complexity of the application rises, this approach begins to run into problems. The main issue is that you have a critical deadline to meet, in that the DSP algorithm must have finished processing the pong buffer before the ADC fills the ping buffer. If, for example, the CPU has to run some other critical code such as user interface or communications stack, then there may not be enough processing power available to run all the necessary codes and the DSP algorithm in time to meet the end of buffer deadline. If the DSP algorithm misses its deadline, then the ADC will start to overwrite the active buffer. You then have to write more sophisticated buffer management code to "catch up," but really you are on to a losing game and data will at some point be lost.

Buffering Techniques: FIFO Message Queue

An alternative to the basic double buffer approach is to use an RTOS to provide a more sophisticated buffering structure using the mailbox system to create a FIFO queue.

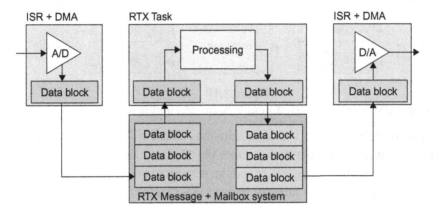

Figure 7.20

Block processing using RTOS mail queues increases the signal latency but provides reliable integration of a DSP thread into a complex application.

Using the RTX RTOS, we can create a task that is used to run the DSP algorithm. The inputs and outputs of this task are a pair of mailboxes. The ADC interrupt is used to fill a mailbox message with a block of ADC data and then post that message to the DSP task. These message blocks are then placed in a FIFO queue ready to be processed. Now if other critical code needs to run, the DSP task will be preempted and rather than risk losing data as in the double buffer case the blocks of data will be sitting in the message queue. Once the critical code has run the DSP, the task will resume running and can use all the CPU processing power to "catch up" with the backlog of data. A similar message queue is used to send blocks of data to the DAC. While you could code this yourself by extending the double buffer example, you would simply end up re inventing the wheel. The RTX RTOS already has message queues with buffer management implemented and as we shall see it results in simple and elegant application code with all the buffer management handled by the RTX RTOS.

In the first example program, the ADC is sampling at 40 KHz and is placed in a 32-byte message buffer.

The message queue is defined by allocating a block of memory as the message space. This region is formatted into a group of mailslots. Each mailslot is a structure in the shape of the data you want to send.

The shape of the mailslot can be any collection of data declared as a structure. In our case, we want an array of floating point values to be our block of data.

```
typedef struct {
float32_t dataBuffer[BLOCK_SIZE];  //declare structure for message queue mailslot
} ADC_RESULT;
```

Next a block of memory is reserved for the mail queue. Here memory is reserved for a queue of 16 mailslots. Data can be written into each mailslot; the mailslot is then locked until the message is received and the buffer released.

```
osMailQDef(mpool_ADC, 16, ADC_RESULT);
```

Then the mailbox can be instantiated by formatting the message box and then initializing the message pointer queue.

```
osMailQId MsgBox;

MsgBox = osMailCreate(osMailQ(MsgBox), NULL);
```

Once this is done, the mailbox is ready to send data between tasks or ISR and tasks. The code below shows the ADC handler logging 32 conversions into a mailslot and then posting it to the DSP routine.

```
void ADC_IRQHandler(void)
{
static ADC_RESULT *mptr;    //declare a mailslot pointer
static unsigned char index = 0;
if(index == 0) {
mptr = (ADC_RESULT*)osMailAlloc(MsgBox, osWaitForever);   //alocate a new mailslot
}
mptr->dataBuffer[index]  = (float)((LPC_ADC->ADGDR>>4)&0xFFF) ;   //place ADC
                                                         data into the mailslot
index++;
if(index == BLOCK_SIZE){
index  = 0;
osMailPut(MsgBox, mptr);   //when mailslot is full send
}
}
```

In the main DSP task, the application code waits for the blocks of data to arrive in the message queue from the ADC. As they arrive, each one is processed and the results are placed in a similar message queue to be fed into the DAC.

```
void DSP_thread (void)
{
ADC_RESULT *ADCptr;     //declare mailslot pointers for ADC and DAC mailboxes
ADC_RESULT *DACptr;
while(1)
{
evt = osMailGet(MsgBox_ADC, osWaitForever);     //wait for a message from the ADC
if (evt.status = = osEventMail)
  ADCptr = (ADC_RESULT*)evt.value.p;
DACptr = (ADC_RESULT*)osMailAlloc(MsgBox, osWaitForever);     //allocate a mailslot for
                                                              the DAC message queue
DSP_Alogrithm  (ADCptr->dataBuffer, DACptr->dataBuffer);    //run the DSP algorithm
osMailPut(MsgBox,DACptr);     //post the results to the DAC message queue
osMailFree(MsgBox, ADCptr);    //release the ADC mailslot
}
}
```

In the example code, a second task is used to read the DAC message and write the processed data to the DAC output register.

```
void DAC_task (void)
{
unsigned int i;
ADC_RESULT *DACrptr;
while(1)
{
evt = osMailGet(MsgBox, osWaitForever);  //wait for a mailslot of processed data
    if (evt.status = = osEventMail)
        DACrptr = (ADC_RESULT*)evt.value.p;
for(i = 0;i < (BLOCK_SIZE);i++)
{
      osSignalWait  (0x01,osWaitForever);  //synchronise with the ADC sample rate
      LPC_DAC->DACR = (unsigned int) DACrptr->dataBuffer[i] << 4;//write a processed
                                                  data value to the DAC register
}
osMailFree(MsgBox, DACrptr);  //free the mailslot
}
}
```

The osSignalWait() RTX call causes the DAC task to halt until another task or ISR sets its event flags to the pattern 0x0001. This is a simple triggering method between tasks that can be used to synchronize the output of the DAC with the sample rate of the ADC. Now if we add the line

```
osSignalSet (tsk_DAC,0x01); //set the event flag in the DAC task.
```

to the ADC ISR, this will trigger the DAC task each time the ADC makes a conversion. The DAC task will wake up each time its event flag is triggered and write an output value to the DAC. This simple method synchronizes the input and output streams of data.

Balancing the Load

The application has two message queues, one from the ADC ISR to the DSP task and one from the DSP task to the DAC task. If we want to build a complex system, say with a GUI interface and a TCP/IP stack, then we must accept that during critical periods the DSP task will be preempted by other tasks running on the system. Provided that the DSP thread can run before the DAC message queue runs dry, then we will not compromise the real-time data. To cope with this, we can use the message queues to provide the necessary buffering to keep the system running during these high demand periods. All we need to do to ensure that this happens is post a number of messages to the DAC message queue before we start running the DAC task. This will introduce additional latency, but there will always be data available for the DAC. It is also possible to use the timeout value in the osMessageGet() function. Here we can set a timeout value to act as a watchdog.

```
evt = osMailGet(MsgBox, WatchdogTimeout);
if(evt.status = = osEventTimeout)
  osThreadSetPriority(t_DSP,osPriorityHigh);
```

Now we will know the maximum period at which the DAC thread must read a data from the message queue. We can set the timeout period to be ¾ of this period. If the message has not arrived, then we can boost the priority of the DSP thread. This will cause the DSP thread to preempt the running thread and process and delayed packets of data. These will then be posted on to the DAC thread. Once the DAC has data to process, it can lower the priority of the DSP thread to resume processing as normal. This way we can guarantee the stream of real-time data in a system with bursts of activity in other threads.

The obvious downside of having a bigger depth of message buffering is the increased signal latency. However, for most applications, this should not be a problem. Particularly when you consider the ease of use of the mailbox combined with the sophisticated memory management of the RTX RTOS.

Exercise: RTX IIR

This exercise implements an IIR filter as an RTX task. A periodic timer interrupt takes data from a 12-bit ADC and builds a block of data that is then posted to the DSP task to be processed. The resulting data is posted back to the timer ISR. As the timer ISR receives processed data packets, it writes an output value per interrupt to the 10-bit DAC. The sampling frequency is 32 KHz and the sample block size is 256 samples.

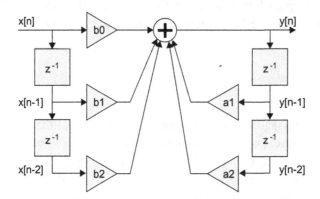

Figure 7.21
An IIR filter is a feedback filter that is much faster than an FIR filter but can be unstable if incorrectly designed.

Open the project in c:\exercises\CMSIS\RTX DSP.

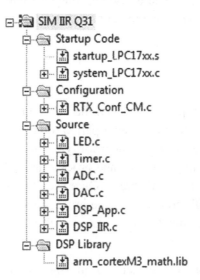

First examine the code, particularly the DSP_App.c module.

The project is actually designed for a Cortex-M3-based microcontroller which as a simulation model that includes an ADC and DAC. The modules ADC.c, Timer.c, and DAC.c contain functions to initialize the microcontroller peripherals. The module DSP_App.c contains the initializing task that sets up the mail queues and creates the DSP task called "sigmod." An additional "clock" task is also created and this task periodically flashes GPIO pins to simulate other activity on the system. DSP_App.c also contains the timer ISR. The timer ISR can be split into two halves. The first section reads the ADC and posts sample data into the DSP task inbound mail queue. The second half receives messages from the outbound DSP task mail queue and writes the sampled data to the DAC. The CMSIS filter functions are in DSP_IIR.c.

Build the project and start the simulator.

When the project starts, it also loads a script file that creates a set of simulation functions. During the simulation, these functions can be accessed via buttons created in the toolbox dialog. If the toolbox dialog is not open when the simulation starts, it can be opened by pressing the "hammer and wrench" icon on the debugger toolbar.

The simulation script generates simulated analog signals linked to the simulator time base and applies them to the simulated ADC input pin.

Open the logic analyzer window.

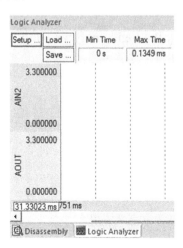

The logic analyzer should have two signals defined, AIN2 and AOUT, each with a range of
0–3.3. These are not program variables but virtual simulation registers that represent the
analog input pin and the DAC output pin.

**Start running the simulator and press the "Mixed Signal Sine" button
in the toolbox dialog.**

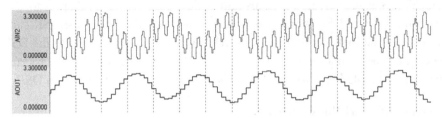

This will generate an input signal that consists of a mixed high- and low-frequency sine wave.
The filter removes the high-frequency component and outputs the processed wave to the DAC.

Now open the Debug\OS Support\Event Viewer window.

In the event viewer window, we can see the activity of each task in the system. Tick the
Cursor box on the event viewer toolbar. Now we can use the red and blue cursors to make
task timing measurements. This allows us to see how the system is working at a task level
and how much processing time is free (i.e., time spent in the idle task).

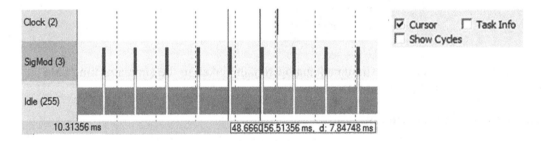

**Stop the simulation and wind back the logic analyzer window to the start of the mixed
sine wave.**

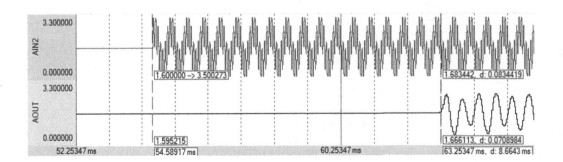

This shows a latency of around 8.6 ms between signal data being sampled and processed data being output from the DAC. While part of this latency is due to the DSP processing time, most of the delay is due to the sample block size used for the DSP algorithm.

Exit the simulator and change the DSP block size from 256 to 1.

The DSP blocksize is defined in IIR.h.

Rebuild and rerun the simulation.

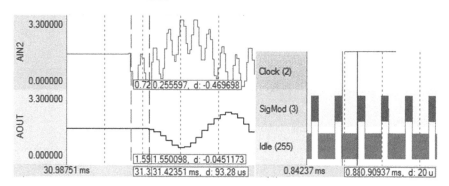

Now the latency between the input and output signal is reduced to around 90 μs. However, if we look at the event viewer, we can see that the SigMod task is running for every sample and is consuming most of the processor resources and is in fact blocking the clock task.

Shouldering the Load, the Direct Memory Access Controller

The Cortex-M3 and Cortex-M4 microcontrollers are designed with a number of parallel internal busses; this is called the "AHB bus matrix lite." The bus matrix allows a Cortex-M-based microcontroller to be designed with multiple bus masters (i.e., a unit capable of initiating a bus access) that can operate in parallel. Typically a Cortex-M microcontroller will have the Cortex processor as a bus master and one or possibly several Direct Memory Access (DMA) units. The DMA units can be programmed by the Cortex-M processor to transfer data from one location in memory to another while the Cortex processor performs another task. Provided that the DMA unit and the Cortex-M processor are not accessing resources on the same arm of the bus matrix, they can run in parallel with no need for bus arbitration. In the case of peripherals, a typical DMA unit can, for example, transfer data from the ADC results register to incremental locations in memory. When working with a peripheral, the DMA unit can be slaved to the peripheral so that the peripheral becomes the flow controller for the DMA transfers. So each time there is an ADC result, the DMA unit transfers the new data to the next location in memory. Going back to our original example by using a DMA unit, we can replace the ADC ISR routine with a DMA transfer that places the ADC results directly into the current mailslot. When the DMA unit has completed its 32 transfers, it will raise an interrupt. This will allow the application code to post the message and configure the DMA unit for its next transfer. The DMA can be used to manage the flows of raw data around the application leaving the Cortex-M processor free to run the DSP processing code.

CHAPTER 8

Debugging with CoreSight

This chapter details the CoreSight debug architecture and in particular its real-time debug features.

Introduction

Going back to the dawn of microcontrollers, development tools were quite primitive. The application code was written in assembler and tested on an erasable programmable read-only memory (EPROM) version of the target microcontroller.

Figure 8.1
The electrical EPROM was the forerunner of today's flash memory.

To help debugging, the code was "instrumented" by adding additional lines of code to write debug information out of the UART or to toggle an IO pin. Monitor debugger programs were developed to run on the target microcontroller and control execution of the application code. While monitor debuggers were a big step forward, they consumed resources on the microcontroller and any bug that was likely to crash the application code would also crash the monitor program just at the point you needed it.

Figure 8.2
An in-circuit emulator provides nonintrusive real-time debug for older embedded microcontrollers.

If you had the money, an alternative was to use an in-circuit emulator. This was a sophisticated piece of hardware that replaced the target microcontroller and allowed full control of the program execution without any intrusion on the CPU runtime or resources. In the late 1990s, the more advanced microcontrollers began to feature various forms of on-chip debug unit. One of the most popular on-chip debug units was specified by the Joint Test Action Group and is known by the initials JTAG. The JTAG debug interface provides a basic debug connection between the microcontroller CPU and the PC debugger via a low-cost hardware interface unit.

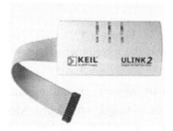

Figure 8.3
Today, a low-cost interface unit allows you to control devices fitted with on-chip debug hardware.

JTAG allows you to start and stop running the CPU. It also allows you to read and write to memory locations and insert instructions into the CPU. This allows the debugger designer to halt the CPU, save the state of the processor, run a series of debug commands and then restore the state of the CPU, and restart execution of the application program. While this process is transparent to the user, it means that the PC debugger program has run control of the CPU (reset, run, halt, and set breakpoint) and memory access (read/write to user memory and peripherals). The key advantage of JTAG is that it provides a core set of debug features with the reliability of an emulator at a much lower cost. The disadvantage of JTAG is that you have to add a hardware socket to the development board and the JTAG interface uses some of the microcontroller pins. Typically, the JTAG interface requires 5 GPIO pins, which may also be multiplexed with other peripherals. More importantly, the JTAG interface needs to halt the CPU before any debug information can be provided to the PC debugger. This run/ stop style of debugging becomes very limited when you are dealing with a real-time system such as a communication protocol or motor control. While the JTAG interface was used on ARM7/9-based microcontrollers, a new debug architecture called CoreSight was introduced by ARM for all the Cortex-M/R- and Cortex-A-based processors.

CoreSight Hardware

When you first look at the datasheet of a Cortex-M-based microcontroller, it is easy to miss the debug features available or assume it has a form of JTAG interface. However, the CoreSight debug architecture provides a very powerful set of debug features that go way beyond what can be offered by JTAG. First of all on the practical side, a basic CoreSight debug connection only requires two pins, serial in and serial out.

Figure 8.4
The CoreSight debug architecture replaces the JTAG berg connector with two styles of subminiature connector.

The JTAG hardware socket is a 20-pin berg connector that often has a bigger footprint on the PCB than the microcontroller that is being debugged. CoreSight specifies two connectors: a 10-pin connector for the standard debug features and a 20-pin connector for the standard debug features and instruction trace. We will talk about trace options later but if your microcontroller supports instruction trace then it is recommended to fit the larger 20-pin socket so you have access to the trace unit even if you do not initially intend to use it. A complex bug can be sorted out in hours with a trace tool where with basic run/stop debugging it could take weeks.

Table 8.1: CoreSight Debug Sockets

Socket	Samtec	Don Connex
10-pin standard debug	FTSH-105-01-L-DV-K	C42-10-B-G-1
20-pin standard + ETM trace	FTSH-110-01-L-DV-K	C42-20-B-G-1

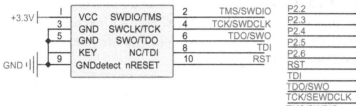 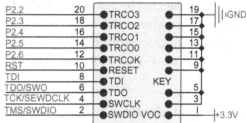

Figure 8.5

The two debug connectors require a minimum number of processor pins for hardware debug.

This standard debug system uses the 10-pin socket. This requires a minimum of two of the microcontroller pins, serial wire IO (SWIO) and serial wire clock (SWCLK), plus the target Vcc, ground, and reset. As we will see below, the Cortex-M3 and Cortex-M4 are fitted with two trace units that require an extra pin called serial wire out (SWO). Some Cortex-M3 and Cortex-M4 devices are fitted with an additional instruction trace unit. This is supported by the 20-pin connector and uses an additional four processor pins for the instruction trace pipe.

Debugger Hardware

Once you have your board fitted with a suitable CoreSight socket you will need a hardware debugger unit that plugs into the socket and is connected to the PC, usually through a USB or Ethernet connection. An increasing number of low-cost development boards also incorporate a USB debug interface. However, many of these use legacy JTAG and do not offer the full range of CoreSight functionality.

CoreSight Debug Architecture

There are several levels of debug support provided over the Cortex-M family. In all cases, the debug system is independent of the CPU and does not use processor resources or runtime.

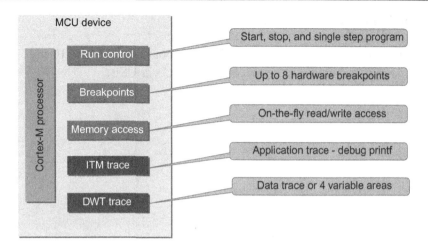

Figure 8.6
In addition to run control the Cortex-M3 and Cortex-M4 basic debug system includes two trace units: a data trace and an instrumentation trace.

The minimal debug system available on the Cortex-M3 and Cortex-M4 consists of the SW interface connected to a debug control system, which consists of a run control unit, breakpoint unit, and memory access unit. The breakpoint unit supports up to eight hardware breakpoints; the total number actually available will depend on the number specified by the silicon manufacturer when the chip is designed. In addition to the debug control units, the standard CoreSight debug support for Cortex-M3 and Cortex-M4 includes two trace units, a data watch trace and an instrumentation trace. The data watch trace allows you to view internal RAM and peripheral locations "on the fly" without using any CPU cycles. The data watch unit allows you to visualize the behavior of your application's data.

Exercise: CoreSight Debug

For most of the examples in this book, I have used the simulator debugger, which is part of the μVision IDE. In this exercise, however, I will run through setting up the debugger to work with the Ulink2 CoreSight debug hardware. While there are a plethora of evaluation boards available for different Cortex devices, the configuration of the hardware debug interface is essentially the same for all boards. There are hardware example sets for the ST Discovery boards, the NXP LPC1768 MBED and the Freescale Freedom boards. The instructions for this exercise customized for each board are included as a PDF in the project directory for each board.

Hardware Configuration

The ULINK2 debugger is connected to the development board through the 10-pin CoreSight debug socket and is in turn connected to the PC via USB.

Figure 8.7
A typical evaluation board provides a JTAG/CoreSight connector. The evaluation board must also have its own power supply. Often this is provided via a USB socket.

It is important to note here that the development board also has its own power supply. The Ulink2 will sink some current into the development board, enough to switch on LEDs on the board and give the appearance that the hardware is working. However, not enough current is provided to allow the board to run reliably; always ensure that the development board is powered by its usual power supply.

Software Configuration

Open the project in exercises\ulink2.

This is a version of the blinky example for the Keil MCBSTM32 evaluation board.

Now open the Options for Target\Debug tab.

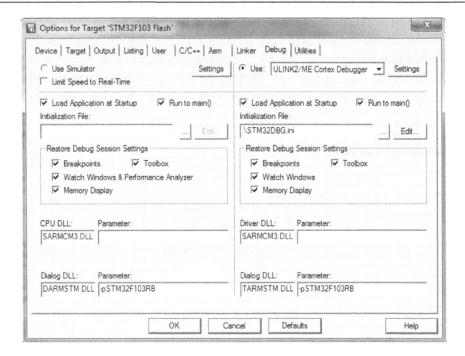

It is possible to switch from using the simulator to the CoreSight debugger by clicking on the Use ULINK2/ME Cortex Debugger radio button on the right-hand side of the menu. Like the simulator it is possible to run a script when the microvision debugger is started. While the debugger will start successfully without this script, the script is used to program a configuration register within the microcontroller debug system that allows us to configure any unique options for the microcontroller in use.

In the case of this microcontroller, the manufacturer has implemented some options to allow some of the user peripherals to be halted when the Cortex CPU is halted by the debug system. In this case, the user timers, watchdogs, and Controller Area Network (CAN) module may be frozen when the CPU is halted. The script also allows us to enable debug support when the Cortex CPU is placed into low-power modes. In this case, the sleep modes designed by the microcontroller manufacturer can be configured to allow the clock source to the CoreSight debug system to keep running while the rest of the microcontroller enters the requested low-power mode. Finally, we can configure the serial trace pin. It must be enabled by selecting TRACE_IOEN with TRACE_MODE set to Asynchronous. When the debugger starts, the script will write the custom config value to the MCU debug register.

```
_WDWORD(0xE0042004, 0x00000027);
```

Now press the Settings button.

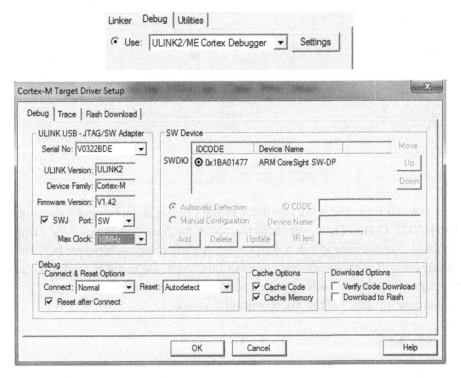

This will open the Ulink2 configuration window. The two main panes in this dialog show the connection state of the ULINK2 to the PC and to the microcontroller. The ULINK2 USB pane shows the ULINK2 serial number and firmware version. It also allows you to select which style of debug interface to use when connecting to the microcontroller. For Cortex-M microcontrollers, you should normally always use SW but JTAG is also available should you need it. The SWJ tick box allows the debugger to select the debug interface style dynamically.

When you are connected to a microcontroller, the SW device dialog will display all of the available debug interfaces (on some microcontrollers there may be more than one). Although you normally do not need to do any configuration to select a debug port, this information is useful in that it tells you that the microcontroller is running. When you are bringing up a new board for the first time it can be useful to check this screen when the board is first connected to get some confidence that the microcontroller is working.

The debug dialog in the settings allows you to control how the debugger connects to the hardware. The Connect option defines if a reset is applied to the microcontroller when the debugger connects; you also have the option to hold the microcontroller in reset. Remember that when you program a code in the flash and reset the processor, the code will run for many cycles before you connect the debugger. If the application code does something to disturb the debug connection such as placing the processor into a sleep mode, then the debugger may fail to connect. The Ulink2 can be configured to hold the Cortex-M processor under reset at the start of a debug session to eliminate such problems. The reset method may also be controlled, and this can be a hardware reset applied to the whole microcontroller or a software reset caused by writing to the SYSRESET or VECTRESET registers in the NVIC. In the case of the SYSRESET option, it is possible to do a "warm" reset, that is, resetting the Cortex CPU without resetting the microcontroller peripherals. The cache options are affected when the physical memory is read and displayed. If the code is cached then the debugger will not read the physical memory but hold a cached version of the program image in the PC memory. If you are writing self-modifying code, you should uncheck this option. A cache of the data memory is also held; if used the debugger data cache is only updated once when the code is halted. If you want to see the peripheral registers update while the code is halted, you should uncheck this option.

Now click on the Trace tab.

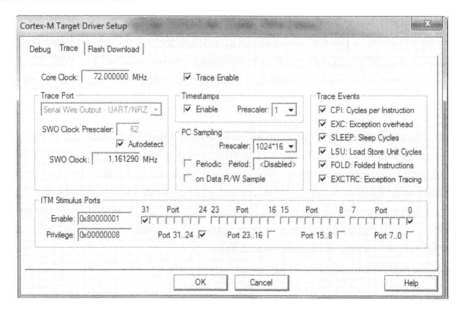

This dialog allows us to configure the internal trace units of the Cortex-M device. To ensure accurate timing, the core clock frequency must be set to the Cortex processor CPU clock frequency. The trace port options are configured for the SW interface using the UART communication protocol. In this menu, we can also enable and configure the data watch trace and enable the various trace event counters. This dialog also configures the instrumentation trace (ITM) and we will have a look at this later in this chapter.

Now click the Flash Download tab.

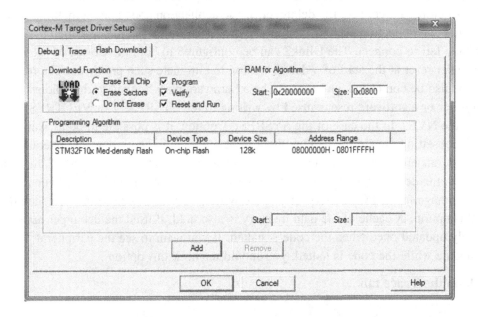

This dialog allows you to set the flash download algorithm for the microcontroller flash memory. This will normally be configured automatically when the project is defined. This menu allows you to update or add algorithms to support additional external parallel or serial flash memory.

Once configured you can start the debugger as normal and it will be connected to the hardware in place of the simulation model. This allows you to exercise the code on the real hardware through the same debugger interface we have been using for the simulator. With the data watch trace enabled you can see the current state of your variables in the watch and memory windows without having to halt the application code. It is also possible to add global variables to the logic analyzer to trace the values of a variable over time. This can be incredibly useful when working with real-time data.

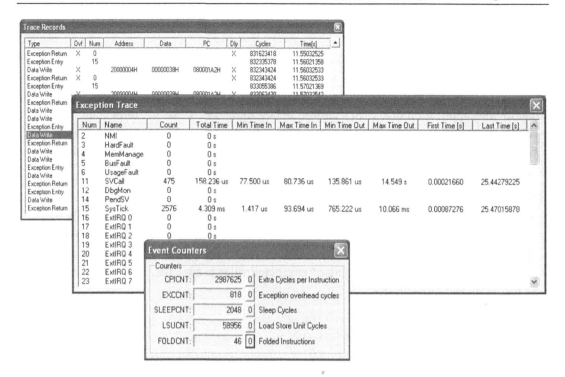

The data watch trace windows give some high-level information about the runtime performance of the application code. The data watch trace windows provide a raw high-level trace exception and data access. An exception trace is also available, which provides detailed information of exception and interrupt behavior. The CoreSight debug architecture also contains a number of counters that show the performance of the Cortex-M processor. The extra cycles per instruction count is the number of wait states the processor has encountered waiting for the instructions to be delivered from the flash memory. This is a good indication at how efficiently the processor is running.

Debug Limitations

When the PC debugger is connected to the real hardware, there are some limitations compared to the simulator. Firstly, you are limited to the number of hardware breakpoints provided by the silicon manufacturer, and there will be a maximum of eight breakpoints. This is not normally a limitation but when you are trying to track down a bug or test code, it is easy to run out. The basic trace units do not provide any instruction trace or timing information. This means that the code coverage and performance analysis features are disabled.

Instrumentation Trace

In addition to the data watch trace, the basic debug structure on the Cortex-M3 and Cortex-M4 includes a second trace unit called the instrumentation trace. You can think of the instrumentation trace as a serial port that is connected to the debugger. You can then add code to your application that writes custom debug messages to the instrumentation trace (ITM) that are then displayed within the debugger. By instrumenting your code this way, you can send complex debug information to the debugger. This can be used to help locate obscure bugs but it is also especially useful for software testing.

The ITM is slightly different from the other two trace units in that it is intrusive on the CPU, i.e., it does use a few CPU cycles. The ITM is best thought of as a debug UART that is connected to a console window in the debugger. To use it, you need to instrument your code by adding simple send and receive hooks. These hooks are part of the CMSIS standard and are automatically defined in the standard microcontroller header file. The hooks consist of three functions and one variable.

```
static __INLINE uint32_t ITM_SendChar (uint32_t ch);    //Send a character to the ITM
volatile int ITM_RxBuffer = ITM_RXBUFFER_EMPTY;         //Receive buffer for the ITM
static __INLINE int ITM_CheckChar (void);               //Check to see if a character has
                                                          been sent from the debugger
static __INLINE int ITM_ReceiveChar (void);             //Read a character from the ITM
```

The ITM is actually a bit more complicated in that it has 32 separate channels. Currently, channel 31 is used by the RTOS kernel to send messages to the debugger for the kernel aware debug windows. Channel 0 is the user channel, which can be used by your application code to send printf() style messages to a console window within the debugger.

Exercise: Setting Up the ITM

In this exercise, we will look at configuring the instrumentation trace to send and receive bytes between the microcontroller and the PC debugger.

Open the project in exercises\CoreSight ITM.

First we need to configure the debugger to enable the application ITM channel.

Open the Options for Target\Debug dialog.

Press the Ulink2 Settings button and select the Trace tab.

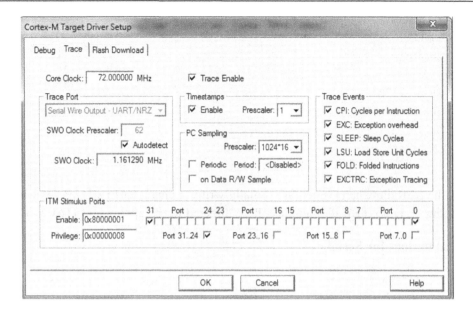

To enable the ITM, the core clock must be set to the correct frequency as discussed in the last exercise and the trace must be enabled.

In the ITM Stimulus Ports menu port 31 will be enabled by default. To enable the application port we must enable port 0. In order to use the application ITM port from privileged and unprivileged mode, the Privilege Port 7..0 box is unchecked.

Once the trace and ITM settings are configured, click OK and return back to the editor.

Now we have to modify the source code to use the ITM in place of a UART.

Open the file Serial.c.

```
void Ser_Init(void);
int SER_PutChar(int c);
int SER_GetChar(void);
int SER_CheckChar(void);
```

This module contains the low-level serial functions that are called by printf() and direct the serial data stream to one of the microcontroller's UARTs. It is also possible to redirect the serial data to the instrumentation trace using the CMSIS Core debug functions.

```
#ifdef __DBG_ITM
volatile int ITM_RxBuffer = ITM_RXBUFFER_EMPTY; /* CMSIS Debug Input */
#endif
```

```
int SER_PutChar (int c) {
#ifdef __DBG_ITM
  ITM_SendChar(c);
#else
  while (!(UART->LSR & 0x20));
  UART->THR = c;
#endif
  return (c);
}
```

For example, the SER_PutChar() function normally writes a character to a UART but if we create the #define __DBG_ITM the output will be directed to the application ITM port.

Open the local options for Serial.c.

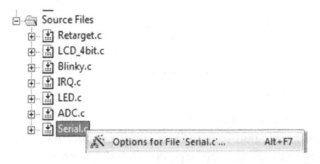

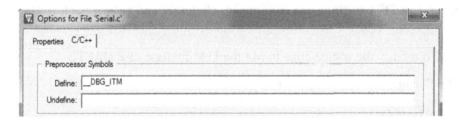

Now add the symbol __DBG_ITM as a local #define.

Close the options menu and rebuild the application code.

Start the hardware debugger.

Open View\Serial Windows\Debug (printf) Viewer.

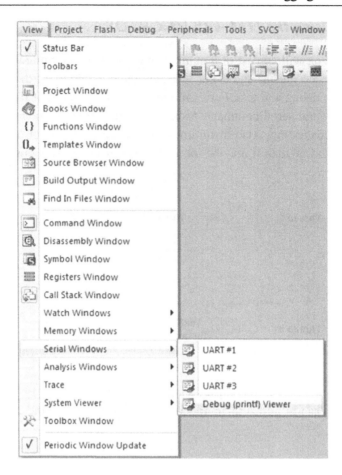

Now start running the code and the serial IO will be redirected to the debugger window rather than the UART.

```
SysTick_Config(SystemCoreClock/100);
printf("Hello World\n\r");
  while (1) {
.........
if (clock_1s) {
  clock_1s = 0;
  printf("AD value: %s\r\n", text);
  }
.........
  }
```

Debug (printf) Viewer
```
Hello World
AD value: 0x01F0
AD value: 0x0DF4
AD value: 0x0FFD
AD value: 0x0FF4
AD value: 0x02CD
AD value: 0x0051
```

It is worth adding ITM support early in your project as it can provide a useful user interface to the running application code. Later in the project, the ITM can be used for software testing.

Software Testing Using the ITM with RTX RTOS

A typical RTX application will consist of a number of threads that communicate via various RTX methods such as main queues and event flags. As the complexity of the application grows, it can be hard to debug and test. For example, to test one region of the code it may be necessary to send several serial command packets to make one task enter a desired state to run the code that needs testing. Also if runtime errors occur we may not pick up on them until the code crashes and by then it may be too late to get meaningful debug information.

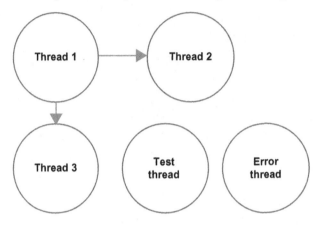

Figure 8.8
Two low-priority threads can be added to an application for error trapping and software testing.

Error Task

In an application, we can introduce two additional threads, an error thread and a test thread. The error thread consists of a switch statement that is triggered by the setting of its signal flags. Each RTX task has 16 error flags. Any other task can set these flags to trigger the task into activity. In the case of the error task, we can use the error flags to send an error code, i.e., a 16-bit number.

```
void errorHandler (void){
uint32_t errorCode;
while(1){
osSignalWait(0xFFFF,osWaitForever);        //wait here until error condition occurs
errorCode = osSignalGet(Error_t);          //read error code
switch(errorCode){                         //switch to run error code code
  case (THREAD2_MAILQUEUE_ERROR)
  ITM_SendChar('[');                        //send an error code to the debugger enclosed in [ ]
  ITM_SendChar('0');
  ITM_SendChar(']');
   break;
 }}}
```

Basically, the task waits until any of its event flags are set. It then reads the state of the error flags to get the "error code." A switch statement then uses the error code to execute some recovery code, and log the error to a file. During development we can also write a message to the debugger via the ITM. So in the main application code we can add statements to trap runtime errors.

```
Error = osMessagePut(Q_LED,0x1,osWaitForever);  //Send a mail message
if(Error! = osOK){  //If the message fails throw an error
  osSignalSetet(t_errorHandler, THREAD2_MAILQUEUE_ERROR);
}
```

In this case, we check for a free message slot before sending a message. If the mail queue is full then the error task is signaled by setting a group of error flags that can be read as an error code. During debug, critical runtime errors can be trapped with this method. Once the code is mature the error task becomes a central location for recovery code and fault logging.

Software Test Task

One of the strengths in using an RTOS is that each of the threads is encapsulated with well-defined inputs and outputs. This allows us to test the behavior of individual threads without having to make changes to the application code, i.e., commenting out functions, creating a test harness, and so on. Instead we can create a test thread. This runs at a low priority so it does not intrude on the overall performance of the application code. Like the error thread, the test thread is a big switch statement but rather than waiting to be triggered by event flags in the application it is waiting for an input from the instrumentation trace.

```
void testTask (void){
unsigned int command;
while(FOREVER){
while (ITM_CheckChar()! = 1) __NOP(); //wait for a character to be sent from the debugger
command = ITM_ReceiveChar();        //read the ITM character
switch(command){                    //switch to the test code
  case ('1'):                       //inject a message into the thread2 mail queue
  test_func_send_thread2_message(CMD_START_MOTOR,0x07,
  0x01,0x07,0,0,0,0);  //send a message to thread2 break;
  case ('2'):                       //halt thread1
  osThreadTerminate(t_thread1);
  default:
  break;
}}}
```

In this case, the test thread is waiting for a character to arrive from the debugger. The switch statement will then execute the desired test code. So for example we could halt a section of the application by deleting the thread, and another debug command could restart it. Then we can inject messages into the RTX mail queues or set event flags to trigger the behavior of the different tasks. This technique is simple and quick to implement and helps to debug application code as it becomes more complex. It is also a basis for the functional testing of mature application code. Because we are using the ITM we only need to access the debug header, and minimal resources are used on the microcontroller. All of the instrumented code is held in one module so it can be easily removed from the project once testing is finished.

Exercise: Software Testing with the ITM

This exercise is designed to run on the MBED and Discovery modules.

Program the project into the microcontroller's flash memory.

Start the debugger and let the code run.

You should see the LEDs on the board flash in a symmetrical pattern.

The application code is located in main_app.c; take a few minutes to look through this to familiarize yourself with this simple application. The CMSIS core layer, which includes the ITM functions, is brought into the project with the microcontroller include file:

```
#include <lpc1768xx.h>
```

The files error.c and test_functions.c provide the error handling and test functions, respectively. For clarity these functions use printf to send test messages to the ITM. In a real application, more lightweight messaging could be used to minimize the function size and number of CPU cycles used. The file Serial.c contains the low-level functions used by printf(). By default these use a standard UART but output can be redirected to the ITM with the #define __DBG_ITM. This is located in the local options for Serial.c.

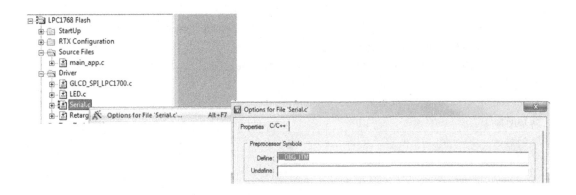

To access the ITM from within the debugger, select View\Serial Windows\Debug (printf) Viewer.

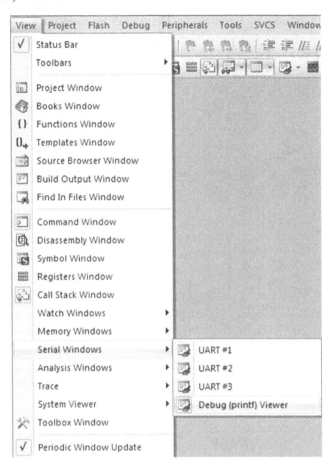

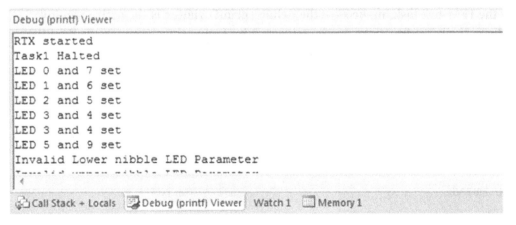

This opens a console style window that allows you to send data to the test task via the ITM and view messages sent from the error task.

You can also view the state of the RTX tasks by opening the Debug\OS Support\RTX Tasks and System window. The event viewer also provides a task trace, which is very useful for seeing the loading on the CPU and the interaction between the different tasks.

ID	Name	Priority	State	Delay	Event Value	Event Mask	Stack Load
255	os_idle_demon	0	Ready				32%
6	testTask	1	Running				0%
5	errorHandler	1	Wait_OR		0x0000	0xFFFF	32%
4	Task3	10	Wait_OR		0x0000	0x0001	32%
2	Task2	10	Wait_MBX				36%
1	Task1	10	Wait_DLY	5			32%

To use the ITM test task, make sure the debug (printf) viewer is the active window. Then the various test functions can be triggered by typing the characters 0–9 on the PC keyboard. The test task provides the following functions.

1—Halt Thread1

2–6—Inject a message into the Thread2 mail queue to switch the LEDs

7—Inject a message with faulty parameters

8—Start Thread1

9—Stop Thread2

0—Start Thread2

Here if you stop Thread2 it will no longer read messages from the mail queue. If Thread1 is running it will continue to write into the mail queue. This will cause the queue to overflow and this will be reported by the error task. Without this kind of error trapping, it is likely that you would end up on the hard fault exception vector and have no idea what had gone wrong with the program. Once the project has been fully developed, the test_functions.c module can be removed and the error task can be modified to take remedial action if an error occurs. This could be logging fault codes to flash, running code to try and recover the error, or worst case resetting the processor.

Instruction Trace with the ETM

The Cortex-M3 and Cortex-M4 may have an optional debug module fitted by the silicon manufacturer when the microcontroller is designed. The ETM is a third trace unit that provides an instruction trace as the application code is executed on the Cortex-M processor. The ETM is normally fitted to higher end Cortex-M3- and Cortex-M4-based microcontrollers. When selecting a device it will be listed as a feature of the microcontroller in the datasheet.

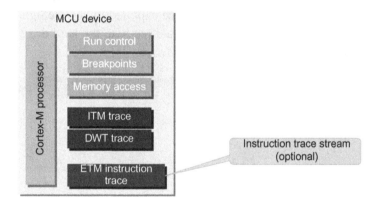

Figure 8.9
The Cortex-M3 and Cortex-M4 may optionally be fitted with a third trace unit. The ETM supports instruction trace. This allows you to quickly find complex bugs. The ETM also enables code coverage and performance analysis tools which are essential for software validation.

The ETM trace pipe requires four additional pins that are brought out to a larger 20-pin socket that incorporates the SW debug pins and the ETM trace pins. Standard JTAG/CoreSight debug tools do not support the ETM trace channel so you will need a more sophisticated debug unit.

Figure 8.10
An ETM trace unit provides all the features of the standard hardware debugger plus instruction trace.

At the beginning of this chapter, we looked at various debug methods that have been used historically with small microcontrollers. For a long time, the only solution that would provide any kind of instruction trace was the in-circuit emulator. The emulator hardware would capture each instruction executed by the microcontroller and store it in an internal trace buffer. When requested, the trace could be displayed within the PC debugger as assembly or a high-level language, typically C. However, the trace buffer had a finite size and it was only possible to capture a portion of the executed code before the trace buffer was full. So while the trace buffer was very useful, it had some serious limitations and took some experience to use correctly. In contrast, the ETM is a streaming trace that outputs compressed trace information. This information can be captured by a CoreSight trace tool; the more sophisticated units will stream the trace data directly to the hard drive of the PC without the need to buffer it within the debugger hardware.

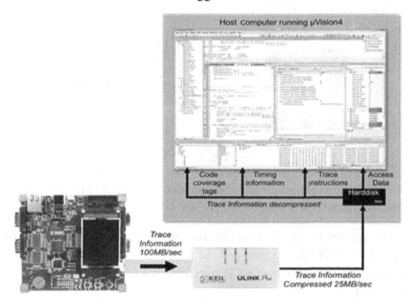

Figure 8.11
A "streaming" trace unit is capable of recording every instruction directly onto the hard drive of the PC. The size of the trace buffer is only limited by the size of your hard disk.

This streaming trace allows the debugger software to display 100% of the instructions executed along with execution times. Along with a trace buffer that is only limited by the size of the PC hard disk it is also possible to analyze the trace information to provide accurate code coverage and performance analysis information.

Exercise: Using the ETM Trace

Open the project in exercises\ETM.

Open the Options for Target\Debug menu.

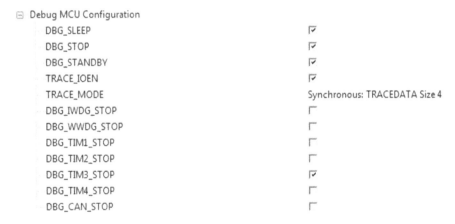

Open the STM32_TP.ini initialization file.

Debug MCU Configuration	
DBG_SLEEP	✔
DBG_STOP	✔
DBG_STANDBY	✔
TRACE_IOEN	✔
TRACE_MODE	Synchronous: TRACEDATA Size 4
DBG_IWDG_STOP	☐
DBG_WWDG_STOP	☐
DBG_TIM1_STOP	☐
DBG_TIM2_STOP	☐
DBG_TIM3_STOP	✔
DBG_TIM4_STOP	☐
DBG_CAN_STOP	☐

This is the same script file that was used with the standard Ulink2 debugger. This time the TRACE_MODE has been set for 4-bit synchronous trace data. This will enable the ETM trace pipe and switch the external microcontroller pins from GPIO to debug pins.

Now press the ULINK Pro Settings button.

Select the Trace tab.

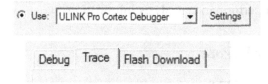

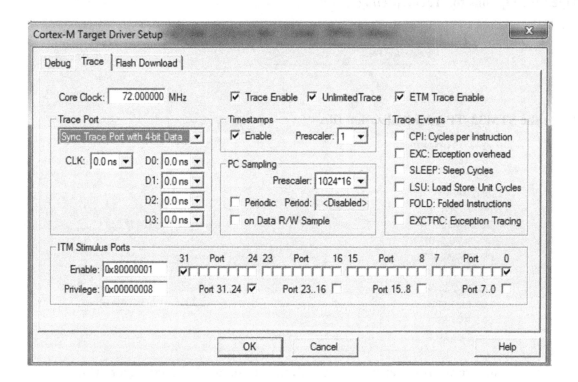

When the ULINK2 trace tool is connected, you have the option to enable the ETM trace. The unlimited trace option allows you to stream every instruction executed to a file on your PC hard disk. The trace buffer is then only limited by the size of the PC hard disk. This makes it possible to trace the executed instructions for days if necessary, yes days.

Click OK to quit back to the μVision editor.

Start the debugger.

Now the debugger has an additional instruction trace window.

Trace Data

Nr.	Time	Address	Opcode	Instruction	Src Code
32,739	0.038 093 547 s	0x00000684	B168	CBZ r0,0x000006A2	
32,740	0.038 093 577 s	0x000006A2	EA950006	EORS r0,r5,r6	if (ad_val ^ ad_val_) { /...
32,741	0.038 093 587 s	0x000006A6	D005	BEQ 0x000006B4	
32,742	0.038 093 617 s	0x000006B4	480D	LDR r0,[pc,#52] ; @0x000006EC	if (clock_1s) {
32,743	0.038 093 637 s	0x000006B6	7800	LDRB r0,[r0,#0x00]	
32,744	0.038 093 657 s	0x000006B8	B130	CBZ r0,0x000006C8	
32,745	0.038 093 687 s	0x000006C8	E7DA	B 0x00000680	while (1) { /* Lo...
32,746	0.038 093 717 s	0x00000680	4815	LDR r0,[pc,#84] ; @0x000006D8	if (AD_done) { /* I...
32,747	0.038 093 737 s	0x00000682	7800	LDRB r0,[r0,#0x00]	
32,748	0.038 093 757 s	0x00000684	B168	CBZ r0,0x000006A2	
32,749	0.038 093 787 s	0x000006A2	EA950006	EORS r0,r5,r6	if (ad_val ^ ad_val_) { /...
32,750	0.038 093 797 s	0x000006A6	D005	BEQ 0x000006B4	
32,751	0.038 093 827 s	0x000006B4	480D	LDR r0,[pc,#52] ; @0x000006EC	if (clock_1s) {
32,752	0.038 093 847 s	0x000006B6	7800	LDRB r0,[r0,#0x00]	
32,753	0.038 093 867 s	0x000006B8	B130	CBZ r0,0x000006C8	

In addition to the trace buffer the ETM also allows us to show the code coverage information that was previously only available in the simulator.

```
60
61        /* AD converter input
62   ⊟    if (AD_done) {
63          AD_done = 0;
64
65          ad_avg += AD_last << 8;
66          ad_avg ++;
67   ⊟      if ((ad_avg & 0xFF) == 0x10) {
68            ad_val = (ad_avg >> 8) >> 4;
69            ad_avg = 0;
70          }
71        }
72
73   ⊟    if (ad_val ^ ad_val_) {
74          ad_val_ = ad_val;
75
76          sprintf(text, "0x%04X", ad_val);
```

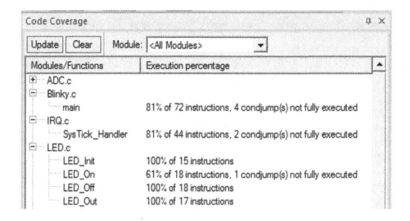

Similarly, timing information can be captured and displayed alongside the C code or as a performance analysis report.

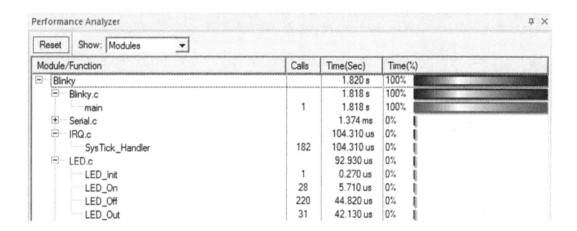

System Control Block Debug Support

The CoreSight debugger interface allows you to control execution of your application code and examine values in the memory and peripheral registers. Combined with the various trace units, this provides you with a powerful debug system for normal program development. However, as we saw in Chapter 3, the Cortex-M processors have up to four fault exceptions that will be triggered if the application code makes incorrect use of the Cortex processor or the microcontroller hardware.

Table 8.2: Fault Exceptions

Fault Exception	Priority	Cortex Processor
Hard fault	−1	Cortex-M0, Cortex-M0+, Cortex-M3, Cortex-M4
Bus fault	Programmable	Cortex-M3, Cortex-M4
Usage fault	Programmable	Cortex-M3, Cortex-M4
Memory manager fault	Programmable	Cortex-M3, Cortex-M4 (Optional)

When this happens your program will be trapped on the default fault handler in the startup code. It can be very hard to work out how you got there. If you have an instruction trace tool, you can work this out in seconds. If you do not have access to instruction trace then resolving a runtime crash can take a long, long time. In this section, we will look at configuring the fault exceptions and then look at how to track back to the source of a fault exception.

The behavior of the fault exceptions can be configured by registers in the system control block. The key registers are listed below.

Table 8.3: Fault Exception Configuration Registers

Register	Processor	Description
Configuration and control	M3, M4	Enable additional fault exception features
System handler control and state	M3, M4	Enable and pending bits for fault exceptions
Configurable fault status register	M3, M4	Detailed fault status bits
Hard fault status	M3, M4	Reports a hard fault or fault escalation
Memory manager fault address	M3, M4	Address of location that caused the memory manager fault
Bus fault address	M3, M4	Address of location that caused the bus fault

When the Cortex-M processor comes out of reset, only the hard fault handler is enabled. If a usage, bus, or memory manager fault is raised and the exception handler for these faults is not enabled, then the fault will "escalate" to a hard fault. The hard fault status register provides two status bits that indicate the source of the hard fault.

Table 8.4: Hard Fault Status Register

Name	Bit	Use
FORCED	30	Reached the hard fault due to fault escalation
VECTTBL	1	Reached the hard fault due to a faulty read of the vector table

The system handler control and state register contains enable, pending, and active bits for the bus usage and memory manager exception handlers. We can also configure the behavior of the fault exceptions with the configuration and control register.

Table 8.5: Configuration and Control Register

Name	Bit	Use
STKALIGN	9	Configures 4- or 8-byte stack alignment
BFHFMIGN	8	Disables data bus faults caused by load and store instructions
DIV_0_TRP	4	Enables a usage fault for divide by zero
UNALIGNTRP	3	Enables a usage fault for unaligned memory access

The divide by zero can be a useful trap to enable, particularly during development. The remaining exceptions should be left disabled unless you have a good reason to switch them on. When a memory manager fault exception occurs, the address of the instruction that attempted to access a prohibited memory region will be stored in the memory fault address register, similarly when a bus fault is raised the address of the instruction that caused the fault will be stored in the bus fault address register. However, under some conditions, it is not always possible to write the fault addresses to these registers. The configurable fault status register contains an extensive set of flags that report the Cortex-M processor error conditions that help you track down the cause of a fault exception.

Tracking Faults

If you have arrived at the hard fault handler, first check the hard fault status register. This will tell you if you have reached the hard fault due to fault escalation or a vector table read error. If there is a fault escalation, next check the system handler control and state register to see which other fault exception is active. The next port of call is the configurable fault status register. This has a wide range of flags that report processor error conditions.

Table 8.6: Configurable Fault Status Register

Name	Bit	Use
DIVBYZERO	25	Divide by zero error
UNALIGNED	24	Unaligned memory access
NOCP	19	No processor present
INVPC	18	Invalid PC load
INVSTATE	17	Illegal access to the execution program status register (EPSR)
UNDEFINSTR	16	Attempted execution of an undefined instruction
BFARVALID	15	Address in bus fault address register is valid
STKERR	12	Bus fault on exception entry stacking
UNSTKERR	11	Bus fault on exception exit on stacking
IMPRECISERR	10	Data bus error. Error address not stacked
PRECISERR	9	Data bus error. Error address stacked
IBUSERR	8	Instruction bus error
MMARVALID	7	Address in the memory manager fault address register is valid
MSTKERR	4	Stacking on exception entry caused a memory manager fault
MUNSTKERR	3	Stacking on exception exit caused a memory manager fault
DACCVIOL	1	Data access violation flag
IACCVIOL	0	Instruction access violation flag

When the processor fault exception is entered, a normal stack frame is pushed onto the stack. In some cases, the bus frame will not be valid and this will be indicated by the flags in the configurable fault status register. When a valid stack frame is pushed, it will contain the PC address of the instruction that generated the fault. By decoding the stack frame, you can retrieve this address and locate the problem instruction. The system control block provides a memory and bus fault address register that depending on the cause of the error may hold the address of the instruction that caused the error exception.

Exercise: Processor Fault Exceptions

Open the project in c:\exercises\fault exception

In this project, we will generate a fault exception and look at how it is handled by the NVIC and how it is possible to trace the fault back to the line of code that caused the problem.

```
volatile uint32_t op1;
int main (void)
{
int op2 = 0x1234,op3 = 0;
  SCB->CCR = 0x0000010;   //enable divide by zero usage fault
  op1 = op2/op3;   //perform a divide by zero to generate a usage exception
while(1);
}
```

The code first enables the divide by zero usage fault and then divides by zero to cause the exception.

Build the code and start the debugger.

Set a breakpoint on the line of code that contains the divide instruction.

```
10 |   SCB->SHCSR = 0x00060000;
11 |   SCB->CCR = 0x0000010;
●12 |   op1 = op2/op3;
```

Run the code until it hits this breakpoint.

Open the Peripherals\Core Peripherals\System Control and Configuration window and check that the divide by zero trap has been enabled.

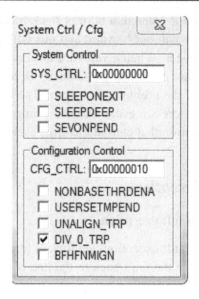

Single step the divide instruction to cause a divide by zero error.

```
128    HardFault_Handler\
129                   PROC
130                   EXPORT   HardFault_Handler          [WEAK]
⇨ 131    |           B        .
132                   ENDP
```

A usage fault exception will be raised. We have not enabled the usage fault exception vector so the fault will elevate to a hard fault.

Open the Peripherals\Core Peripherals\Fault Reports window.

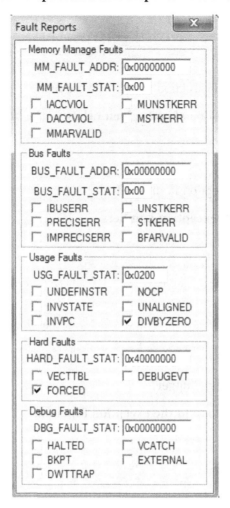

This window shows that the hard fault has been forced by another fault exception. Also the divide by zero flag has been in the usage fault status register.

In the register window, read the contents of R13, the main stack pointer, and open a memory window at this location.

Read the PC value saved in the stack frame and open the disassembly window at this location.

```
0x08000198 FB91F2F0  SDIV   r2,r1,r0
0x0800019C 4B08      LDR    r3,[pc,#32]  ; @0x080001C0
```

This takes us back to the SDIV instruction that caused the fault.

Exit the debugger and add the line of code given below to the beginning of the program.

```
SCB->SHCSR = 0x00060000;
```

This enables the usage fault exception in the NVIC.

Now add a C level usage fault exception handler.

```
void UsageFault_Handler (void)
{
  error_address = (uint32_t *)(__get_MSP());  // load the current base address of the
stack pointer
  error_address = error_address + 6;  // locate the PC value in the last stack frame
  while(1);
}
```

Build the project and start the debugger.

Set a breakpoint on the while loop in the exception function.

```
19    void UsageFault_Handler (void)
20 ⊟ {
21        error_address = (uint32_t *)(__get_MSP());
22        error_address = error_address + 6;
●23       while(1);
24    }
```

Run the code until the exception is raised and the breakpoint is reached.

When a usage fault occurs this exception routine will be triggered. It reads the value stored in the stack pointer and extracts the value of the PC stored in the stack frame.

CMSIS SVD

The CMSIS SVD format is designed to provide silicon manufacturers a method of creating a description of the peripheral registers in their microcontrollers. This description can be passed to third party tool manufacturers so that compiler includes files and debugger peripheral view windows can be created automatically. This means that there will be no lag in software development support when a new family of devices is released. As a developer you will not normally need to work with these files but it is useful to understand how the process works so that you can fix any errors that may inevitably occur. It is also possible to create your own additional peripheral debug windows. This would allow you to create a view of an external memory mapped peripheral or provide a debug view of a complex memory object.

When the silicon manufacturer develops a new microcontroller, they also create an XML description of the microcontroller registers. A conversion utility is then used to create a binary version of the file that is used by the debugger to automatically create the peripheral debug windows.

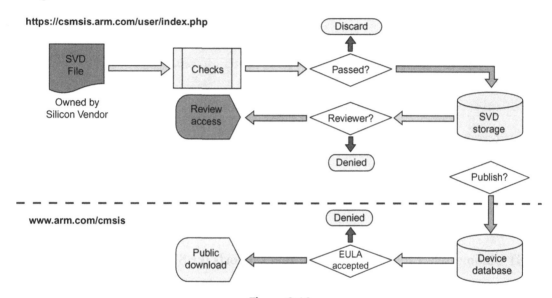

Figure 8.12
ARM has created a repository of "SVD" files. This enables tools suppliers to have support for new devices as they are released.

Alongside the XML description definition, ARM has introduced a submission and publishing system for new SVD files. When a silicon manufacturer designs a new microcontroller, its system description file is submitted via the CMSIS Web site. Once it has been reviewed and validated, it is then published for public download on the main ARM Web site.

Exercise: CMSIS SVD

In this exercise, we will take a look at a typical system viewer file to make an addition and to rebuild the debugger file, and then check the updated version in the debugger.

For this exercise, you will need an XML editor. If you do not have one, then download a trial or free tool from the Internet.

Open your web browser and go to www.arm.com/cmsis.

If you do not have a user account on the ARM Web site, you will need to create one and login.

On the CMSIS page, select the CMSIS-SVD tab.

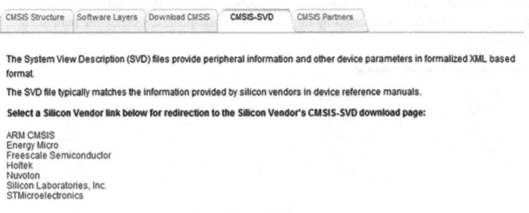

Figure 8.13
The CMSIS Web site has public links to the current CMSIS specification and the CMSIS-SVD repository.

Select the STMicroelectronics link.

In the ST window, select the SVD download that supports the STM32F103RB.

Then click the download button at the bottom of the page.

Unzip the compressed file into a directory.

Open the STM32F103xx.svd file with the XML editor.

Each of the peripheral windows is structured as a series of XML tags that can be edited or a new peripheral pane can be added. This would allow you to display the registers of an external peripheral interfaced onto an external bus. Each peripheral window starts with a name, description, and group name followed by the base address of the peripheral.

Once the peripheral details have been defined, a description of each register in the peripheral is added. This consists of the register name and description of its offset from the peripheral base address along with its size in bits, access type, and reset value.

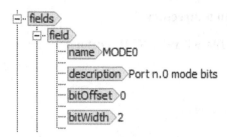

It is also possible to define bit fields within the register. This allows the debugger window to expand the register view and display the contents of the bit field.

Make a small change to the XML file and save the results.

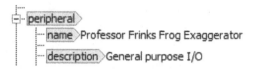

Generate the SFR file by using the SVDConv.exe utility.

This utility is located in c:\keil\arm\cmsis\svd.

The convertor must be invoked with the following command line:

SVDConv STM32F103xx.svd–generate = sfr

This will create STM32F103xx.SFR.

Now start μVision and open a previous exercise.

Open the Options for Target menu and select STM32F103xx.sfr as the system viewer file.

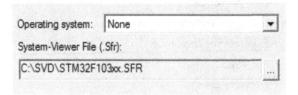

Build the project and start the debugger.

Open the View\System Viewer selection and view the updated peripheral window.

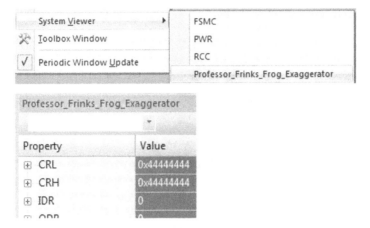

End of book craziness aside, it is useful to know how to add and modify the peripheral windows so you can correct mistakes or add in your own specific debug support.

CMSIS DAP

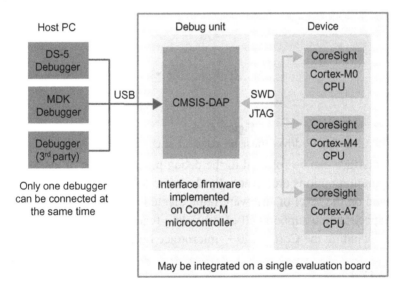

Figure 8.14

The CMSIS DAP specification is designed to support interoperability between different debugger hardware and debugger software.

The CMSIS DAP specification defines the interface protocol between the CoreSight debugger hardware and the PC debugger software. This creates a new level of interoperability between different vendors' software and hardware debuggers. The CMSIS DAP firmware is designed to operate on even the most basic debugger hardware. This allows even the most basic evaluation modules to host a common debug interface that can be used with any CMSIS-compliant tool chain.

Figure 8.15
The MDED module is the first to support the CMSIS DAP specification.

The CMSIS DAP specification is designed to support a USB interface between the target hardware and the PC. This allows many simple modules to be powered directly from the PC USB port.

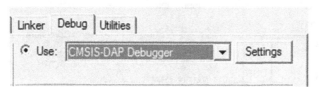

Figure 8.16
The CMSIS DAP driver must be selected in the debugger menu.

The CMSIS DAP interface can be selected in the debug menu in place of the proprietary Ulink2. The configuration options are essentially the same as the Ulink2 but the options available will depend on the level of firmware implemented by the device manufacturer. The CMSIS DAP specification supports all of the debug features found in the CoreSight debug architecture including the Cortex-M0+ microtrace buffer (MTB).

Cortex-M0+ MTB

While the ETM is available for the Cortex-M3 and Cortex-M4, no form of instruction trace is currently available for the Cortex-M0. However, the Cortex-M0+ has a simple form of instruction trace buffer called the MTB. The MTB uses a region of internal SRAM that is allocated by the developer. When the application code is running, a trace of executed instructions is recorded into this region. When the code is halted, the debugger can read the MTB trace data and display the executed instructions. The MTB trace RAM can be configured as a circular buffer or a one-shot recording. While this is a very limited trace, the circular buffer will allow you to see "what just happened" before the code halted. The one-shot mode can be triggered by the hardware breakpoints to start and stop allowing you to track down more elusive bugs.

Exercise: MTB

This exercise is based on the Freescale Freedom board for the MKL25Z microcontroller. This was the first microcontroller available to use the Cortex-M0+.

Connect the Freedom board via its USB cable to the PC.

Open the project in c:\exercises\CMSIS DAP.

Open the Options for Target\Debug menu.

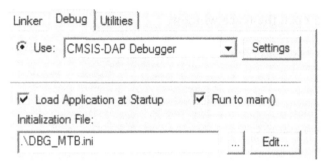

Here the CMSIS DAP interface is selected along with an initializing file for the MTB.

The initializing script file has a wizard that allows you to configure the size and MTB.

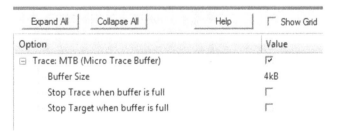

Here we can select the amount of internal SRAM that is to be used for the trace. It is also possible to configure different debugger actions when the trace is full: either halt trace recording or halt execution on the target.

By default the MTB is located at the start of the internal SRAM (0x20000000). So it is necessary to offset the start of user SRAM by the size of memory allocated to the MTB. This is done in the Options for Target\Target dialog.

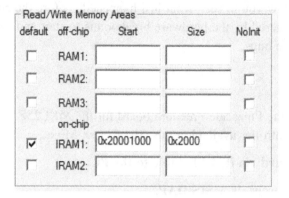

Now start the debugger and execute the code.

Halt the debugger and open the trace window.

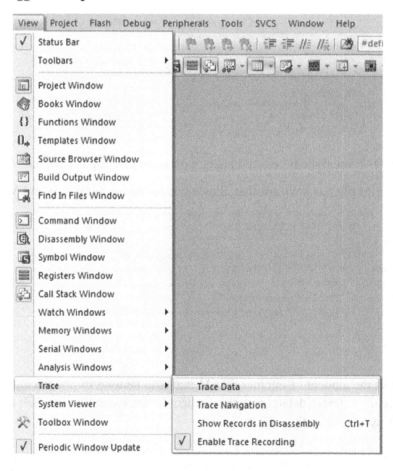

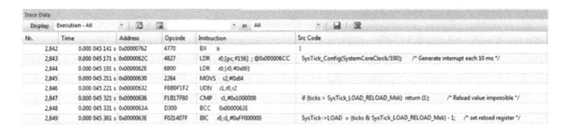

While this is a limited trace buffer, it can be used by very low-cost tools and provides a means of tracking down runtime bugs that would be time consuming to find any other way.

Debug Features Summary

Table 8.7: Summary of Cortex-M Debug Features

Feature	Cortex-M0 Cortex-M0+	Cortex-M3 Cortex-M4
Debug interface	Legacy JTAG or SW	Legacy JTAG or SW
"On the fly" memory access	Yes	Yes
Hardware breakpoint	4	6 Instruction + 2 Literal
Data watchpoint	2	4
Software breakpoint	Yes	Yes
ETM instruction trace	No	Yes (optional)
Data trace	No	Yes (optional)
Instrumentation trace	No	Yes
SW viewer	No	Yes
MTB	Yes (Cortex-M0+ only)	No

Appendix

This appendix lists some further resources that are worth investigating once you have worked through this book

Debug Tools and Software

There are lots of development tools available for ARM Cortex-M microcontrollers. The following is by no means an exclusive list. I have just tried to list some of the key software resources available.

Commercial GNU-Based Toolchains

Atollic	www.atollic.com
Rowley	www.rowley.co.uk
Code Red	http://www.code-red-tech.com/

Commercial Toolchains

IAR	www.iar.com
Keil	www.keil.com
Tasking	www.tasking.com

Auto Coding

National Instruments	http://www.ni.com/labview/arm/

Rapid Prototyping and Evaluation

Mbed	mbed.org

Mbed is a low-cost Cortex-M-based module with a web-based development environment. It has a thriving community with lots of free resources.

Operating Systems

Like the compiler toolchains, there are many RTOS for Cortex-M processor. Here I have listed some of the most widely used RTOS vendors.

Free Rtos	www.freertos.org
Keil	www.keil.com
Micrium	Micrium.com
Segger	www.segger.com
CMX	www.cmx.com

DSP

DSP resources	www.dspguru.com
Filter Design	www.iowegian.com
DSP algorithm development	www.mathworks.co.uk

Other Tools

CANdroid	Mobile data and Signal Analysis for your Smartphone	www.hitex.co.uk/index.php?id=3333
PC Lint	The essential static analysis tool also supports MISRA C	www.gimple.com
Powerscale	Energy profiling tool	www.hitex.co.uk/index.php?id= powerscale-system

Books

Programming

The C Programming Language Kernighan and Ritchie.

Cortex-M Processor

Cortex-M Technical Reference Manuals	http://www.arm.com/products/processors/ cortex-m/index.php
The Definitive guide to the Cortex M3 Joseph Yui	
Insiders Guide to the STM32 Martin	http://www.hitex.co.uk/index.php?id= download-insiders-guides00

CMSIS

CMSIS Specification	www.onARM.com
MISRA C	www.misra.org.uk
Barr Coding standard	http://www.barrgroup.com/ Coding-StandardResources

RTOS

Real-Time Concepts for Embedded Systems	Qing Li and Caroline Yao	
Little Book of Semaphores	Downey	http://www.greenteapress.com/ semaphores/

DSP

Digital Signal Processing: A Practical Guide for Engineers and Scientists Stephen Smith	
Understanding Digital Signal Processing	Lyons

Silicon Vendors

There are now over 1000 Cortex-M-based microcontrollers and the list is still growing. The link below is a database of devices supported by the MDK-ARM. This is an up-to-date list of all the mainstream devices and is a good place to start when selecting a device.

Cortex-M-device database	http://www.keil.com/dd/

Training

The following companies provide hands on Embedded systems training courses:

Hitex.co.uk	http://www.hitex.co.uk/ index.php?id=3431
Feabhas	www.Feabhas.com
Doulos	www.Doulos.com

Accreditation

ARM maintains an accreditation program which provides Professional Certification for Development Engineers working with ARM.

http://www.arm.com/support/arm-accredited-engineer/index.php

Contact Details

If you have any questions or comments, please contact me on the e-mail below:

Cortex.Designer@hitex.co.uk

You can also register for a monthly Embedded systems Newsletter.

http://news.hitex.co.uk/newsletter/4PZ-3QW/signup.aspx

Index

Note: Page numbers followed by "*f*" and "*t*" refer to figures and tables, respectively.

Printed in the United States
By Bookmasters